GOLD PANNING the Pacific Northwest

A Guide to the Area's Best Sites for Gold

Second Edition

GARRET ROMAINE

ESSEX, CONNECTICUT

FALCONGUIDES®

An imprint of Globe Pequot, the trade division of
The Rowman & Littlefield Publishing Group, Inc.
4501 Forbes Blvd., Ste. 200
Lanham, MD 20706
www.rowman.com

Falcon and FalconGuides are registered trademarks and Make Adventure Your Story is a trademark of The Rowman & Littlefield Publishing Group, Inc.

Distributed by NATIONAL BOOK NETWORK

Photos by Garret Romaine
Maps by The Rowman & Littlefield Publishing Group, Inc.

British Library Cataloguing in Publication Information available

Library of Congress Cataloging-in-Publication Data
Names: Romaine, Garret, author.
Title: Gold panning the Pacific Northwest : a guide to the area's best sites for gold / Garret Romaine.
Description: Second edition. | Essex, Connecticut : FalconGuides, [2023] | Includes bibliographical references and index. | Summary: "A reference for readers interested in getting started or continuing their gold prospecting in the Pacific Northwest region. Fully revised and updated, this guide contains accurate, up-to-date prospecting information for all known panning areas in Oregon, Washington, and Idaho. The write-ups for each locale include driving directions, GPS coordinates, historical information, land ownership restrictions, full-color photos, and geological background"— Provided by publisher.
Identifiers: LCCN 2022049072 (print) | LCCN 2022049073 (ebook) | ISBN 9781493064434 (paper ; alk paper) | ISBN 9781493064441 (electronic)
Subjects: LCSH: Gold panning—Northwest, Pacific. | Gold mines and mining—Northwest, Pacific. | Northwest, Pacific—Guidebooks.
Classification: LCC TN423.A5 R66 2023 (print) | LCC TN423.A5 (ebook) | DDC 622/.342209795—dc23/eng/20221018
LC record available at https://lccn.loc.gov/2022049072
LC ebook record available at https://lccn.loc.gov/2022049073

∞™ The paper used in this publication meets the minimum requirements of American National Standard for Information Sciences—Permanence of Paper for Printed Library Materials, ANSI/NISO Z39.48-1992.

To my cousin John—a great addition to any gold trip, by Crom!

CONTENTS

Part II: Oregon

ACKNOWLEDGMENTS

I've been scouting out gold-panning trips around the Pacific Northwest since the 1970s. My dad and uncle would load up a pickup truck on Friday night and head out for the weekend to eastern Oregon on what they called "sod trips," returning to their favorite old haunts where they grew up. One time we panned concentrates from the old E&E Mine near Bourne with a tin plate; another time we found intriguing residues in an old ball mill at the Badger Mine near Susanville and brought them home in a coffee can. Ever since then, it has been a fascinating journey to far-flung old camps and abandoned mines and a quest for ever-better equipment.

Along the way these recon trips have involved a lot of old and new friends, and it's been fun to teach newcomers how to pan or identify rocks and show off some of our favorite haunts. I know there are still a lot of folks that I would like to take on a trip, and I owe at least one local club a guided tour to Bohemia City. Here's a genuine "Recon Salute" to the many folks

Crumbling ruins at Mineral, Idaho

who helped me out in the field: the late Val Bailey and his son Josh; Tom Bohmker; Deb Chen; Pferron Doss; the late Dan Driscoll; Tim and Tonya Fisher; Frank Higgins; Josh Krause, Ellen Schippers, and their son Wyatt; Paul Leiker; Nathan O'Brien; Ely Perry; Kevan Reedy; Martin Schippers; Dirk Williams; Jake and Kyle Riley; Terry Snyder; Nick Henson; and Eric and Ford Veeder. I'm sure I missed a few people, and for that I apologize.

Special thanks go to my family: my cousin John and his son Dougie Romaine; my late uncle Doug Romaine and his wife, Janet; my late father, Garret L. Romaine; my son, Nelson Romaine; my daughter, Amelia Perry Romaine; and most of all, my long-suffering wife, Cindy Romaine, who was there that fateful night when we got two flats at once and left the Boise National Forest on a flatbed truck headed for the nearest Les Schwab Tire Center. She tells the story better.

PREFACE

This book is aimed at everyone who has had even a small case of gold fever. If you have ever seen a small pinpoint of gold emerge from the black sands and concentrates at the bottom of a gold pan, you know what I mean. If you have never experienced the thrill of recovering an elusive piece of gold, you have a major emotion in front of you. There are very few things that quicken the pulse like seeing actual gold gleam from pay dirt.

Since I was a kid in the 1970s, I have traveled regularly to the historic mining districts in this region, primarily as a columnist for *Gold Prospectors* magazine and before that as a senior editor at *North American Gold Mining Industry News*. I have learned where to go, places to avoid, and what to find. I cannot guarantee to make you rich, because this is not a get-rich scheme. It's a labor of love, a lot of fun, and a ticket to some of the most interesting places out in the hills. What I can do is save you time, effort, and money as you research places to explore.

Over the years, I have developed a deep nostalgia for the gold rushes and mining frenzies of the Old West. It is such a thrill to drive up a long, winding mountain road, turn the corner, and see the remains of an old mining camp come into view. Hiking to abandoned mines is another adventure I wish I could share—there is a true sense of accomplishment when you reach the end of a trail and find that old mine shaft. Sometimes you can close your eyes and imagine hundreds, if not thousands, of noisy miners and merchants in those old settings, only to be interrupted by the call of a stern old turkey or an irritated squirrel. I have watched many old mills and head frames crumble into piles of rotted timbers, but the tailings piles still yield great hand specimens. Every spring fresh torrents move and mix the sands and gravels anew and re-sort the pay streaks. There is still a lot of gold out there, if you have the time and energy to go about recovering it. Most of the easy stuff is gone, but that just makes the thrill of finding what's there that much stronger.

By reading this book, you will save yourself a lot of trouble out in the wilder parts of the region, and you will benefit from some of my worst mistakes. I have seen gates spring up across once-reliable access roads, but I've also watched GPS and Google Earth become major forces for preventing confusion in the field. I have collected hundreds of GPS coordinates for you to use as aids, and I have personally visited the areas that I cover here. Whether you

Nice shiny gold in brilliant white quartz

have an old gold pan under the seat of your truck or a new Gold Cube packed away, you will benefit from this book.

The old Oregon Country is the name given to the drainage of the Snake River, covering Washington, Oregon, Idaho, and parts of Wyoming and Montana. As a proud member of the Geological Society of the Oregon Country, I'm going to err on the side of providing you with as much geology as I feel is prudent. I am going to explain host rocks that contain quartz veins, and I'm going to name the granite batholiths that brought so much prosperity to this region. I'm going to try to educate you about collecting rock samples, and I'm going to follow that up by discussing the common modern machines that grizzled old prospectors would have killed for back in the day. By the end of this book, you'll find gold, you'll learn some history, and you'll have a great appreciation for the gold rush days that settled this region. Most of all, you will visit some of the storied districts that gave this region much of its resilience and vigor. You'll visit places that you'll want to talk about with your friends and family, and you'll see wildlife, landforms, and vistas that will stir your soul. I hope you enjoy your search.

INTRODUCTION

Of all the legends and stories about gold, the one that is hardest to explain is the adage "Gold is where you find it." The old saying seems to imply that if you look hard enough just about anywhere, you can stumble upon a nugget or two. Nothing could be further from the truth. A better way to modernize this would be to point out that "Gold is where they found it." The reason is pretty simple—the gold prospectors that came before us were really, really good at what they did. For all the images conjured up of a slightly addled Yosemite Sam with his burro and pick, there have been a lot of smart geologists since then that scientifically sampled every major gold district. Many mines started production, exhausted sizable ore deposits, and then disappeared. The likelihood of discovering a new district is extremely remote.

The good news is that, while they were efficient, the miners of the late 1800s were also in a hurry. They creamed off all the easy gold—the concentrated pockets, the big nuggets, and the "potato diggings" monsters went to the smelter early. Their primitive equipment skipped a lot of fine gold, too, but they couldn't get it all. Some machines simply washed the "flour" gold over the transom, out the end of the dredge, to remix with the rest of the tailings. Miners resorted to mercury to collect some of the fines, but very few miners use mercury anymore in the Pacific Northwest. It is toxic and dangerous, and modern equipment can do just as well, if not better, at recovering fine gold. Which is good, because that's a lot of what's left.

If there is one scientific fact you can count on, it is that it will always be impossible to get all the gold from any particular area. There will always be a little color left, deep in the crevices, under the big boulders, or lying on hard clay layers known as false bedrock.

A dedicated weekend hobbyist, recreational prospector, or otherwise-motivated gold miner can still go one of two routes in the search for gold:

Hard-rock mining involves dynamite, cyanide leaching, crushers, big rigs, and lots of capital. There are still fortunes to be had out there with hard-rock mining, but the odds are long. Throughout most of the West, any sizable deposit has likely been identified and claimed up. You could still get lucky, but it would take a lot of work. You could rework old tailings and do well, or you could find an area thought to yield copper that has good gold left. Some districts, such as Cornucopia, Oregon, were shut down at the beginning of

Shiny gold flakes and nuggets from Pacific Northwest locales

World War II and never restarted. It would take millions of dollars now to get those mines restarted. Hard-rock mining is capital intensive, and it involves a lot of machines.

Placer mining involves equipment ranging from a simple gold pan all the way up to trucks, excavators, and a wash plant, as you've no doubt seen on Friday night's *Gold Rush* series on the Discovery Channel. This type of gold mining usually involves less investment and lode mining and will consistently yield small amounts of gold, with occasional bonanzas for those who are persistent or willing to throw a lot of money at the task. If you can learn to reliably return from every trip with decent concentrates, so that over time you fill a five-gallon bucket, and then maybe even a fifty-five-gallon drum, with black sands, magnetite, ilmenite, rare earth elements (REEs), and gold, you will be rewarded in the long run.

Either way, your long-term goals are your own. Very few prospectors are simply in it for the money, looking at this as a way to become a millionaire overnight. Some of us just like to get out of town, camp in the mountains, and enjoy the spirit of the outdoors. Some people like to work up a little sweat and appetite, improve their health, and learn a little. Some of us like to solve problems and run machinery and enjoy the challenge of keeping a pump going or making sure the sluice is running right. Still others like the wildlife, the scenery, and the historical importance of the Wild West, and bring back their riches as photos and videos. In each case, if you toss in a little gold fever as motivation and stay scientific about your sampling and exploration, you will prosper far and above the value of your recovered material.

Still, a nice payday is always a treat. One sure way to reach that goal is to keep trying. Keep practicing, keep exploring, and keep getting out in the field. Another truism that seems to hold is that the farther away from civilization you get, the better your chances. That holds for the coordinates in this book as well. Find a spot, look around, and see what else is out there.

HOW TO USE THIS GUIDE

The best way to use this guide is to treat it like a general reference for the multistate area. Here are some likely scenarios.

You might live near a district listed here or have always wanted to visit it. Based on what you read and learn, you will have more information to process, and then you can decide if you want to explore that area further. Thanks to the GPS coordinates found in these pages, you can easily pinpoint a few spots that you will be able to drive right to, thus maximizing your time in the field so that you are efficient and productive.

If you already have scheduled a business trip, family vacation, or other reason for driving through an area, you might check in these pages and determine that you will be close enough to a spot that is worth more exploration. The information here can help you minimize the detour, making it easier to justify. In addition, if you need to camp, I have listed developed and primitive camping options.

If you are considering a major investment in mining supplies, I have some good information for you in appendix A. Over the course of writing nearly one hundred articles for *Gold Prospectors*, the magazine of the Gold Prospectors Association of America (GPAA), I have reviewed many products. I can provide you with good, practical information about metal detectors, dredges, highbankers, sluices, and more. Note that I stuck that information at the back of the book so that you can get right to the good stuff: where to go.

If you are just getting started in the hobby, I recommend picking up *The Modern Rockhounding and Prospecting Handbook*, a companion book I wrote for FalconGuides. The information there can make you a better field geologist by explaining the basics of economic ore deposits, field sampling, and identifying hand specimens. You will get good overview information about the basics there. If you need help identifying rocks and minerals, I wrote rock-and-gem identification books for several regions as well (see appendix B).

In addition to the GPS coordinates for these locales, I have also included road directions. I have to warn you though: Conditions change fast. US Forest Service (USFS) or Bureau of Land Management (BLM) offices may schedule major roadwork during your prized vacation window. Floods, washouts, fires, and other road damage can leave you sputtering in front of a hand-lettered sign at two o'clock in the morning, facing a 4-hour detour. It is important

Tumbleweeds are taking over this old homestead in central Washington.

that you take the road directions I provide with a healthy dose of caution and either phone the local government agency managing that land or consult its web page if you don't have time to track down a human. The more you are counting on taking your family and friends to a single location for a long stay, the more important it is that you check in ahead of time.

Note: Roads in national forests have the prefix NF, and BLM roads use a BLM-# format.

Map Legend

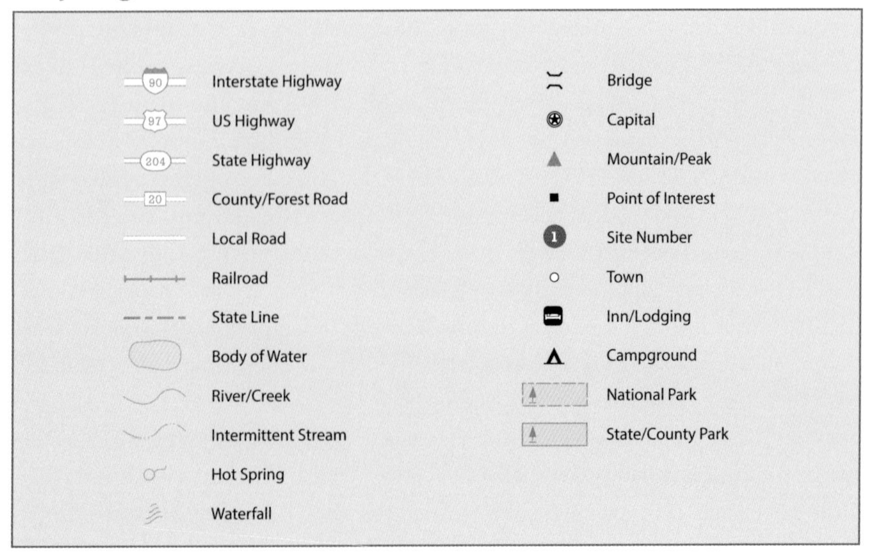

Part I: Washington

Washington

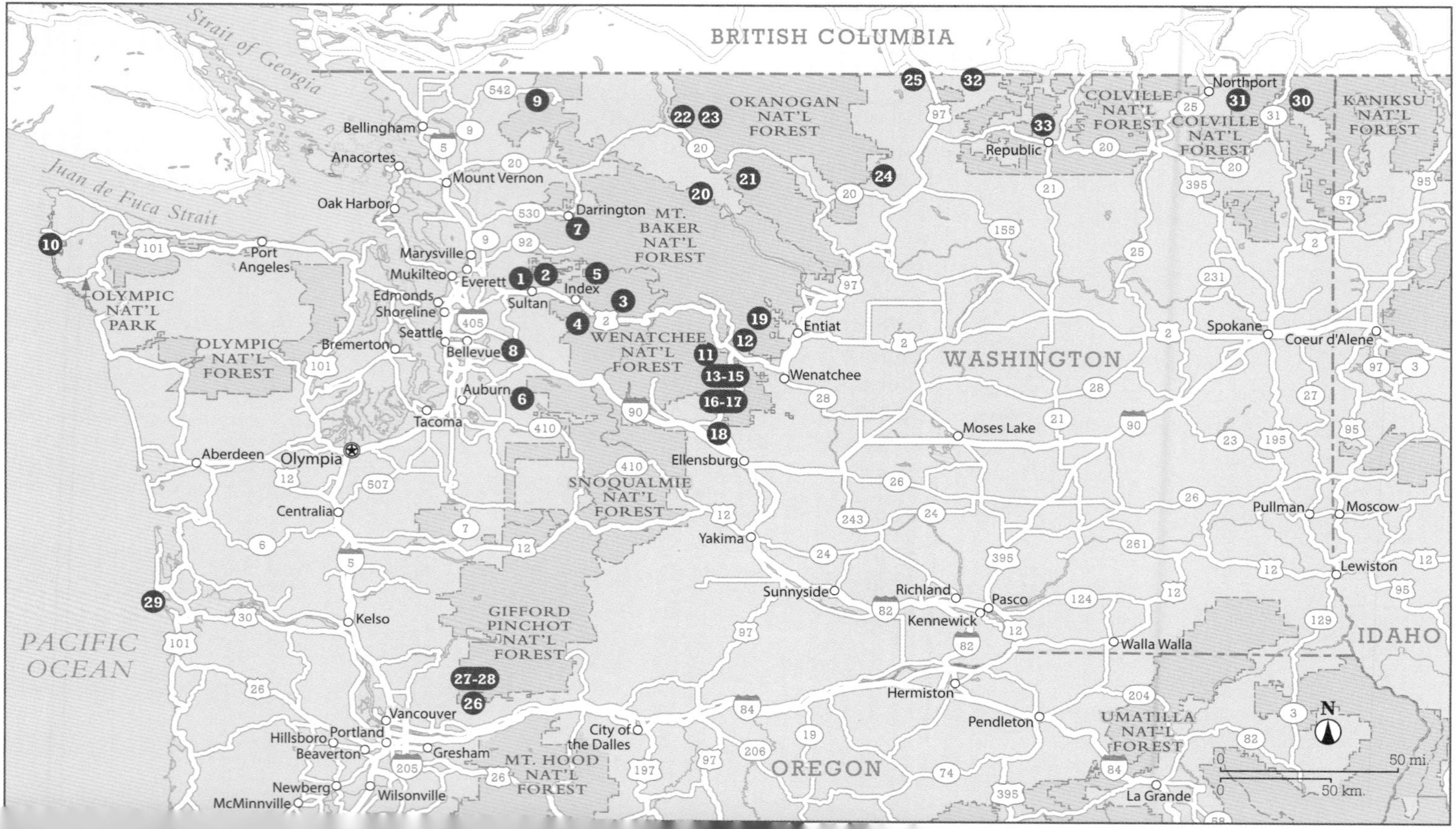

BRITISH COLUMBIA
Strait of Georgia
Juan de Fuca Strait
OKANOGAN NAT'L FOREST
COLVILLE NAT'L FOREST
KANIKSU NAT'L FOREST
MT. BAKER NAT'L FOREST
OLYMPIC NAT'L PARK
OLYMPIC NAT'L FOREST
WENATCHEE NAT'L FOREST
SNOQUALMIE NAT'L FOREST
GIFFORD PINCHOT NAT'L FOREST
MT. HOOD NAT'L FOREST
UMATILLA NAT'L FOREST
WASHINGTON
IDAHO
OREGON
PACIFIC OCEAN
Bellingham
Anacortes
Mount Vernon
Oak Harbor
Port Angeles
Marysville
Mukilteo
Everett
Edmonds
Shoreline
Seattle
Bremerton
Bellevue
Sultan
Index
Darrington
Auburn
Tacoma
Olympia
Aberdeen
Centralia
Kelso
Vancouver
Portland
Hillsboro
Beaverton
Gresham
Newberg
Wilsonville
McMinnville
City of the Dalles
Entiat
Wenatchee
Ellensburg
Yakima
Sunnyside
Richland
Kennewick
Pasco
Hermiston
Pendleton
Walla Walla
La Grande
Moses Lake
Republic
Northport
Spokane
Coeur d'Alene
Pullman
Moscow
Lewiston
N
0 50 mi.
0 50 km.

NORTHWEST WASHINGTON

1 Sultan River

Land type: Creek, riverbank
County: Snohomish
GPS:
A - Sportsman Park: 47.86175, -121.82209
B - Mouth of river: 47.86043, -121.82023
C - Sultan Placer: 47.85995, -121.81777
D - Fishing spot: 47.84507, -121.92669

Northwest Washington

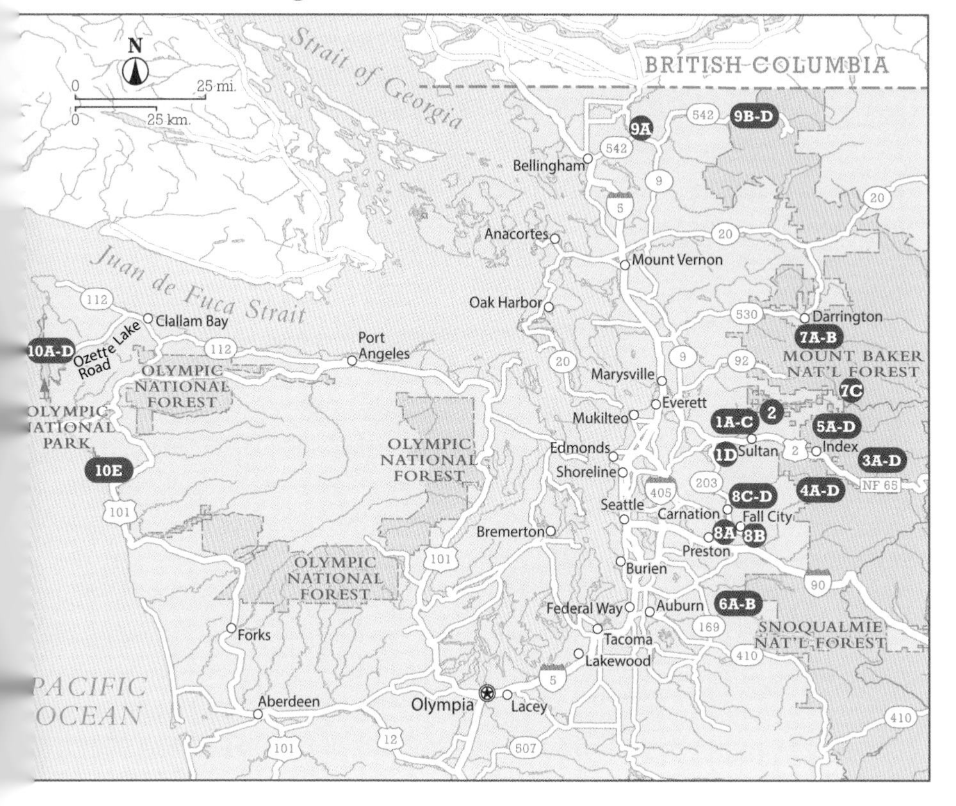

Best season: Late summer
Land manager: County owns park; state owns land along river
Material: Fine flood gold
Tools: Pan, wheel, dredge with permit (in season)
Vehicle: Any
Special attractions: GPAA Gold Show at Monroe, annual Gold Dust festival in Gold Bar
Accommodations: Motels and RV camping in Sultan; camping at Wallace Falls State Park; dispersed camping on USFS land at Money Creek
Finding the sites: From I-5 head east to Monroe. You can come in via WA 522 or US 2. From Monroe drive east about 7.5 miles. As you come into Sultan, look for Sportsman Park on the north side of the highway and the west side of the Sultan River. The Skykomish River is south. Drive into the park and get your bearings; look for the spot at the bend in the Sultan as it rounds the park. There are additional access points along the west bank. To reach the mouth of the river, locate the path under the highway and the railroad. To reach the Sultan Placer, drive east on US 2 for 0.4 mile, turn right (south), cross the Skykomish, go about 0.25 mile total from US 2, and look for a parking area to the west. This is a fishing access point; you want the trail that leads back across the road and to the west. You can see the mouth of the Sultan River on the far bank if you find the right place. The farthest spot west is another public fishing spot; access it via Lewis Street/WA 203 from Monroe, go south across the Skykomish River, then turn left onto Ben Howard Road and drive 2.1 miles.

Prospecting

When the water is low, look for bedrock that you can scrape or dig against. If the river is too high, you can also scrape moss and wash it in a five-gallon bucket, then pan those sands for flood or flour gold. You have more of a chance of getting a flake if you can find bedrock, but you need to come here in late summer. The county park has a nice bend in the river here, but it is a long way to bedrock. You can easily reach the mouth of the river from Sportsman Park. There were several placer mines along this stretch of the Skykomish River, including the Jonkers, Phoenix, Alda, and Ida Mines. The mines were on both sides of the river, with the best pay streaks on the bottom. Note that there is good gold on the Sultan River above the park, but the power company now locks up the Horseshoe Bend site, and the Washington Prospectors Mining Association (WPMA) controls the rest of the river. Check

This bedrock crops out at the public fishing spot below Sultan.

them out at washingtonprospectors.org. If you join this organization, you can get behind the locked gate!

Remember that any time you are gold panning in Washington, you must have the "Gold and Fish" pamphlet, written by the Department of Natural Resources. Here's the link: https://wdfw.wa.gov/licenses/environmental/hpa/types/prospecting.

2 Olney Creek

See map on page 7.
Land type: Creek bank
County: Snohomish
GPS: Bridge: 47.92368, -121.74353
Best season: Summer–Oct
Land manager: State forest
Material: Fine gold, some flakes; garnets
Tools: Pan, sluice, wheel; pry bar for big rocks
Vehicle: Any; paved road
Special attraction: Spada Lake
Accommodations: Camping OK; no facilities, so pack out your trash.
Finding the site: About 1 mile east of Sultan, turn north onto Sultan Basin Road. Stay on the road for 7.2 miles until you find the bridge across Olney Creek. Park safely. To the north of the road, you should see some good gravel showings. You can continue another 3 miles northeast, with Olney Creek below and to your left.

Olney Creek is still accessible in late October, but it's cold.

Prospecting

There are good small flakes here if you can get low enough. This spot is popular and has been included in mining club catalogs in the past, but by late 2022 the entire creek was open and free of claims. That gives you plenty of room to roam here and locate friendly terrain for gold to settle around big rocks and on inside bends of the creek. There are big cobbles in the gravels, made from interesting metamorphic rock such as schist and gneiss. You will find plenty of attractive red garnets in the concentrates as well. Try starting a hole along the side of the creek on an inside bend and check as you go; you should notice the gold getting bigger as you dig farther down. You can explore the creek in both directions, and the farther you get from the bridge, the better you are likely to do. There are camping spots as well to the northeast. Consult a map before heading too far south, although prospectors report values all the way to the mouth of Olney Creek; stay within earshot of the road if possible. Be sure to fill in any big holes you create, and pick up any garbage left by hunters. Don't be surprised to hear gunshots, as hunters like to sight in their rifles out here.

3 Beckler River

See map on page 7.
Land type: River
County: Snohomish
GPS:
A - Bridge: 47.72739, -121.33788
B - Beckler River CG: 47.73363, -121.33333
C - Pull-out: 47.79957, -121.29329
D - Junction: 47.80283, -121.29274
Best season: Summer–early fall
Land manager: Mount Baker–Snoqualmie National Forest
Material: Fine gold, black sands; garnets
Tools: Pan, sluice, dredge
Vehicle: Any; 4WD suggested
Special attraction: Stevens Pass
Accommodations: Primitive camping all along Beckler River; developed camp at Beckler River CG
Finding the sites: Drive east on US 2 from Monroe to Skykomish, about 35 miles. There are services in Skykomish. About 0.8 mile past Skykomish, look for a left turn (north) onto Beckler River Road. The bridge locale is about a mile from the highway, and the campground is about 1.5 miles from the highway. You should find a decent pull-out to access the river at 6.7 miles. Finally, the upper place to access the Beckler River, at the junction with NF 6530, is about 6.9 miles from US 2. Someday, the Beckler River Road may once again offer a loop all the way to Index. In late 2022 there was construction at the washout. You can at least access the North Fork of the Skykomish River and the Galena/Mineral City district, but there are so many club claims up there that I didn't include the locale here.

Prospecting

We always tend to get sidetracked on the garnets up here, but the Olympic Placer was a big operation back in the 1930s, and the area is noted for gold as well. Look for fine gold, plenty of black sands, rare fragments of dark staurolite crystals, and plentiful garnets, some big enough to roll around the pan like buckshot. If you have the patience, you can extract quite a pile of reddish

Bring a pry bar to move the big boulders at the Beckler River.

garnets and show them off in a glass jar. Barring that attraction, you will want to move boulders and get low in the riverbank to reach pay streaks. If you do not have much time and just want to sample the river, you should do fine at the bridge locale at Site A; scout out a nice beach downstream from the bridge and work around the bedrock exposed there. The Beckler River Campground at Site B sits right on the river, and we had fun there one year setting up a panning wheel on a picnic table. The former Olympic Placer near the mouth of Johnson Creek is claimed, so it has been removed from this edition. The next 4 miles of river access are unclaimed (though you might spot very old primitive claim signs here and there) until you near Jack Pass. The garnets may clog up your riffles here, but you should see small pieces of gold in your pans, especially if you can get a hole going all the way to bedrock. The pull-out at Site C is another good access spot, but it's a plan B in case the primitive camping area at the junction is overpopulated.

4 Money Creek

See map on page 7.
Land type: Creek
County: King
GPS:
A - Bridge: 47.71633, -121.40995
B - Easy access: 47.70869, -121.43341
C - More access: 47.69838, -121.47208
D - Damon & Pythias Mine: 47.69887, -121.52721
Best season: Late spring–early fall
Land manager: Mount Baker–Snoqualmie National Forest
Material: Fine gold, black sands
Tools: Pan, sluice, dredge
Vehicle: Any; 4WD suggested
Special attraction: Stevens Pass
Accommodations: Primitive camping all along Money Creek and Miller River; developed camp at Money Creek CG
Finding the sites: From the town of Grotto, drive east on US 2 about 0.8 mile and turn right (south) onto the old highway. Drive about 1.1 miles and turn right onto Miller River Road. After less than 0.1 mile, turn right onto Money Creek Road. After 0.6 mile you reach the bridge; this is Site A. NF 6422 continues on the left, but you want Money Creek Road/NF 6420, which goes somewhat right. Begin prospecting at Site B, about 1.2 miles from the bridge. You will find many more contact points along the creek for the next couple miles; there is easy access about 3.3 miles from the bridge, at Site C. At 6.1 miles take NF 6420 straight for another 0.3 mile to NF 610, which is now a footpath that covers about 0.2 mile to the mouth of the Damon & Pythias Mine.

Prospecting

Money Creek is another famed locale below Stevens Pass where Depression-era miners did well. There are big boulders just below the bridge here that trap larger pieces of gold but require more work to move. The bridge marks the eastern edge of section 29, and there are two valid claims west of the bridge. If you just want a quick sample, the rocks along this creek on the east side

Tailings pile at the Damon & Pythias Mine.

of the bridge hide under green moss, which acts as a kind of "miner's moss" and traps black sands and gold. Scrape the moss and the sand layer under it into a bucket, wash thoroughly, and pan the concentrates. This is a sampling technique; do not go crazy scalping all the boulders. Otherwise, you can work up Money Creek, the Miller River, and even the nearby Foss River for several miles, although I did not find much color in either drainage. There are many hard-rock mines in this area, and Money Creek held multiple placer claims and mines in its day. The famed Damon & Pythias Mine is still a valid claim, but the tailings piles hold interesting pyrite samples if you dig into them. This mine is a worthwhile field trip in summer, sponsored by the Washington State Mineral Council, and highly recommended. The nearby Apex Mine resides about 0.75 mile due southeast, up the very steep mountainside, at 47.6955, -121.5123.

5 Index

See map on page 7.
Land type: River
County: Snohomish
GPS:
A - Boat launch: 47.83644, -121.65915
B - Line A Placer: 47.81622, -121.56696
C - Trout Creek: 47.86323, -121.48692
D - Washout: 47.86345, -121.48238
Best season: Spring–fall
Land manager: Mount Baker–Snoqualmie National Forest
Material: Fine gold, small flakes, black sands; garnets
Tools: Pan, sluice, highbanker, dredge
Vehicle: Any; paved roads
Special attraction: Index
Accommodations: Camping at Wallace Falls State Park and at Line A; dispersed camping near washout; avoid private land between washout and Index.
Finding the sites: From Gold Bar on US 2, drive east 2.2 miles to the bridge over the Skykomish River. Cross the river, drive 0.15 mile, and look for access to the recreation area on the right. To reach the primitive camp at the old Line A placer mine, a once-valid claim now lapsed, drive a total of 7.4 miles from Gold Bar to the Index-Galena Road, which bears left. After 0.9 mile take the bridge across the river into Index and stay on Fifth Street until it reaches Index Avenue. Go left and proceed about 0.3 mile, then go left again on Second Street. Now go right onto Avenue A and drive along the river about 0.4 mile to the parking area. To reach Site C at Trout Creek, go back to Index, cross over the river, and head up the Index-Galena Road for 6.1 miles. Park safely and work your way along Trout Creek to the river. The washout is another 0.2 mile from Trout Creek, but as of 2022 there was construction there, hopefully to restore access to Galena and Mineral City.

Prospecting

Your concentrates up here will be rich in black sands and tiny garnets, so be aware that riffles could clog if sluicing. The Skykomish River Boat Ramp is a good picnic area, where author Tom Bohmker (2009) reported nice colors

Once the road crews repair this washout above Index, you will have easier access to more spots on the North Fork of the Skykomish River.

around the boulders about 150 feet upstream from the foot of the boat ramp, which we confirmed in 2014. As always, fill in your holes. At the old Line A placer mine below Index, you can find large boulders on an inside bend, plus bedrock—an unbeatable combination, especially with camping just feet away. This is a popular spot that fills up fast on summer weekends. Farther up the Index-Galena Road, you can use Trout Creek to steer you to a good spot on the river, with an excellent bar forming downstream from the mouth of the creek. The Sunset Mine trailhead nearby offers access to the mine and is an interesting hike. I listed the washout more for reference here—it is far from the main stem of the river. The river washed out the road in 2006, and there is finally a plan in place to fix it, which will make access to Galena, Silver Creek, and Mineral City much easier. As it stands right now, you have to go up the Beckler River to Jack Pass and turn left, down the hill, and access the S-bridge area at Galena that way. Prospectors Plus has a large block of claims on the Skykomish below the bridge, and Washington Prospectors has a great locale above the bridge, closer to the mouth of Silver Creek. I interviewed some local miners up here, and they reported that there is probably too much private land up at Mineral City to make a hike all the way up there worthwhile.

6 Green River

See map on page 7.
Land type: Creek
County: King
GPS:
A - Bridge: 47.31913, -121.89346
B - Access: 47.31265, -121.87212
Best season: Spring–early fall; avoid salmon run, as river smells awful then.
Land manager: Washington Dept. of Natural Resources/King County
Material: Fine gold
Tools: Pan, sluice
Vehicle: Any; paved roads
Special attraction: Green River Gorge
Accommodations: Camping at Kanaskat-Palmer State Park; motels at Maple Valley
Finding the sites: From Auburn drive west on WA 18 about 6.9 miles to WA 516/ SE 272nd and turn right. Go 5.1 miles to WA 169, where WA 516 turns into Kent Kangley Road, and take it another 3.4 miles. Go right onto Retreat-Kanaskat Road, drive 3.1 miles, then veer right onto Cumberland Kanaskat Road. Drive 1.2 miles to the bridge. To reach the upper spots, drive 0.3 mile to Palmer, turn left onto SE Green River Headworks Road, and go 1.1 miles to Pipeline Road.

Prospecting

This area offers some close-in gravels to prospect if you live in the Seattle area. It would not be worth planning an entire summer around, but there really was a large placer mine at the bend in the river (Site A) in the 1930s. Mayo (1983) reports a Depression-era miner worked the gravel near Palmer and made wages, and "a piece of rough quartz showing wire gold was . . . found by a fisherman . . . near the Kummer Bridge" (p. 24). Good luck finding the source; the roads are all private, and Howard A. Hanson Reservoir covers many acres. Many locals use the beaches at the bridge for recreation, so you will probably see vehicles there as you get close. You can park safely and work your way down to a nice beach with plenty of large boulders and bedrock to explore; there are leaf fossils in the dark shale material. You will

This bend in the Green River was the site of a minor placer operation back in the Depression era.

need low water to have much success here, but it's nice to know this river has something to offer besides its grim history as a dumping ground for Gary Ridgway, the notorious "Green River Killer." Prospectors continue to report large flakes and small nuggets from the river between here and the end of public access. One caution: Avoid this area when the salmon run is under way, as the shores reek from abundant carcasses by mid-October, and the smell is all but unbearable.

7 Darrington

See map on page 7.
Land type: Riverbank, creek
County: Snohomish
GPS:
A - Squire Creek CG: 48.27092, -121.67499
B - Clear Creek CG: 48.21916, -121.56878
C - Monte Cristo: 48.01254, -121.44208
Best season: Late summer
Land manager: Mount Baker–Snoqualmie National Forest
Material: Fine gold, small flakes, black sands; garnets
Tools: Pan, sluice, shovel, bucket, hammer
Vehicle: Any; 4WD suggested
Special attraction: Asbestos Falls
Accommodations: Squire Creek, Clear Creek, and Bedal CGs; dispersed camping along the Sauk River above Whitechuck
Finding the sites: From the intersection of WA 530 and Mountain Loop Highway in downtown Darrington, drive west on WA 530 for 3.6 miles on the Mountain Loop. Turn north into Squire Creek Campground, and work your way to the west of the campsites. You can reach the creek via a short walk from several spots. To reach Clear Creek CG, again from the intersection in Darrington, follow the Mountain Loop Highway south about 3 miles to the campground, or continue a bit farther to the bridge and park safely there. To reach Barlow Pass, drive from Darrington on Mountain Loop Highway about 23 miles. Park safely and hike in toward Monte Cristo about a mile.

Prospecting

The Darrington area saw quite a bit of placer and lode gold mining in the late 1800s and again during the Great Depression. Squire Creek was a good producer, but a lot of the best mineralization is locked up in the Boulder River Wilderness. There is only about 0.5 mile of open creek to explore between private land and the wilderness, and it is far too challenging to keep in for this second edition. The best area to work Squire Creek is way up the fork where Buckeye Creek comes in from the west, but the Squire Creek Campground at

The South Fork of the Sauk River drains the area around Monte Cristo, seen here above Barlow Pass.

Site A is interesting if you need a place to camp anyway. At the mouth of Clear Creek at Site B, there is a solid outcrop of bedrock right below the bridge, and the river is constantly washing the gravels and leaving black sands and fine gold behind. There are some large boulders to move as well, and if you are in a hurry, you can clean some moss off the rocks.

Up at Monte Cristo there was extensive mineralization in the hills above the old townsite. Back in the day, Monte Cristo boasted a decent little rush, and Frederick Trump, grandfather of the 45th president of the United States, was a Bavarian immigrant who opened a hotel (and brothel) here before following the gold rush to Alaska. It's a shame the road washed out, but the 4- to 5-mile hike is worth it, as it follows the old road. If you make a weekend of it up here, be sure to hike all the way to the townsite, but be respectful of the buildings. There is a valid placer claim near the Gothic Basin trailhead, but most of the upper Sauk River is open. Look for obvious bends in the river and prepare to do a little scouting before committing to a dig site. Also, be sure to fill in your holes. There are old placer diggings throughout this area, especially past Monte Cristo on Glacier Creek, and an old trail leading to Poodle Dog Pass that will connect you to Mineral City, Galena, and Index if you've got the energy. It has been said to be a treacherous passage that requires ice axes, crampons, and pitons. Consult the Northwest Underground Explorations series for more information about the lode mines here—a big hammer will be worth it on the many tailings piles.

8 Raging River

See map on page 7.
Land type: River
County: King
GPS:
A - Raging River: 47.51786, -121.92448
B - Zerfleuh boat launch: 47.56787, -121.88309
C - Tolt River 1: 47.63869, -121.92645
D - Tolt River 2: 47.63829, -121.91769
Best season: Late summer
Land manager: Mount Baker–Snoqualmie National Forest
Material: Fine gold, small flakes, black sands
Tools: Pan, sluice, shovel, bucket
Vehicle: Any
Special attraction: Snoqualmie Falls
Accommodations: Twin Falls State Park; Tinkham CG; dispersed camping along South Fork of the Snoqualmie River
Finding the sites: To reach the Raging River site, head out of Seattle on I-90 east to exit 22. Cross back over the freeway and turn right onto SE High Point Way. Follow it for 0.5 mile, then turn right onto Upper Preston Road SE. Drive 0.5 mile and look for ample parking on the right. To reach the Zerfleuh boat launch, which is on the South Fork of the Snoqualmie River, backtrack on Upper Preston Road SE, past more access to the Raging River, and return to Preston-Fall City Road. Go right (north) for 3.7 miles, turn right onto Dike Road, then look for the park. To reach the Tolt River sites, go back to Dike Road, then drive north on Fall City-Carnation Road/WA 503S from the boat launch about 5.2 miles, before the bridge, and turn left onto NE Tolt Hill Road. Go about 0.5 mile and park. This is Tolt River 1. To get to the second Tolt River site, go back to Fall City-Carnation Road, turn left, and cross the bridge. In 0.1 mile make a sharp turn to the left down to the parking area.

Prospecting

Many of the rivers east of the Seattle metro area contain traces of gold, including the Cedar, Mad, Tolt, and Raging Rivers, according to Mayo (1983). This Raging River site is intriguing—there is a nice inside bend here, it is easy

This spot on the Raging River is directly beneath the I-90 overpass.

to get to, and there are ample black sands and decent colors. The source is a mystery—perhaps it is from an old channel of the Snoqualmie River. You *must not* dig around the interstate highway bridge pilings, tempting as that might be at first. Move some rocks around and get a big hole going, and you should do fine. Just be sure to fill that hole back in. If you have the right permits, you could probably run a highbanker here. There are three parks near Fall City that provide access to the area where the Raging River dumps into the Snoqualmie, but I liked the Zurfleuh boat launch site best. Make your way to the mouth and note the large gravel bar below when the water is low. I would stick to panning only here. There was a former placer mine on the upper Tolt River at 47.7053, -121.8051. It is behind a gate now. Still, the Tolt River has good black sands and fine, flour gold available for panning. There is decent public access on both sides of the mouth of the Tolt River. Even the black sands are fine here, but there are a few accumulations of fist-size cobbles that you can dig out and wash off.

9 Glacier

See map on page 7.
Land type: River, creek
County: Whatcom
GPS:
A - Nugents Corner: 48.84064, -122.29214
B - Silver Fir CG: 48.90549, -121.82271
C - Open cut: 48.90525, -121.80804
D - Great Excelsior Mine: 48.89879, -121.80607
Best season: Late summer
Land manager: Mount Baker–Snoqualmie National Forest
Material: Fine gold, small flakes, black sands
Tools: Pan, sluice, shovel, bucket, hammer
Vehicle: Any
Special attraction: Nooksack Falls
Accommodations: Douglas Fir and Silver Fir CG; dispersed camping along the Nooksack River, Glacier Creek Road, and Canyon Creek Road
Finding the sites: Nugents Corner is the lowest spot, reachable from I-5 via exit 255 onto WA 542. Drive 9.9 miles toward Nugents Corner, crossing the bridge, and then look for a sharp right turn. Drive back to the river and park. To reach the sites up by Glacier, continue east on WA 542 to just past Deming, where you pick up WA 9 and head north; after 12.9 miles, you will reach Kendall. Turn right (east) and continue on WA 542 for about 17.2 miles to the turn for Silver Fir Campground.
To reach the site near Nooksack Falls, continue east on WA 542 from the turn for Silver Fir Campground for 0.7 mile, then take a slight right onto NF 33. Wind down the hill for about 0.5 mile to parking. Enjoy the falls, then cross the bridge and hike about 0.1 mile to the old adit, a brilliant orange and yellow. If you continue on about 0.5 mile, there is a sharp turn to the left to an old greenstone quarry at 48.9023, -121.8045. There is more mineralization another 0.5 mile up the road.
To reach the Great Excelsior Mine, turn right off WA 542 about 1.1 miles east of Glacier onto Glacier Creek Road/NF 39. After 100 feet, swing left onto NF 3. After about a mile you'll come to a road heading northwest that leads to the old Glacier Mine. You can explore the dumps there for copper ore; malachite is easy to spot, but gold was only a by-product here, so skip it if you are short on time. The trailhead for the Great Excelsior Mine is a total of 6.5 miles from WA 542 via NF 3.

Inspecting the tailings and rubble at the Great Excelsior Mine

The GPS coordinates are 48.9034, -121.8142. Follow the trail about 0.6 mile as it gets worse and worse; stay on it until you reach the mine.

Prospecting

The panning down at Nugents Corner takes patience—look on the inside bend of the river and search for the area where larger rocks are piling up. The gold tends to trap under and downstream from individual rocks. You may have to fill a bucket and take it down to the water to pan. Always fill in your holes, but here especially. Closer to Glacier you will find all kinds of interesting rockhounding. There is high-grade coal out of Glacier Creek and ancient fossils in the hills. Northwest Underground Explorations did a great job documenting this area as well. The panning at Silver Fir Campground is fair—you are much closer to the source of the mineralization, mostly coming down Wells Creek, so expect larger pieces and coarse black sands. The open cut is a good example of the ore material in this part of the Cascades. It's not the glamorous visible gold on white quartz made famous in California's 16-to-1 Mine—here you'll find microscopic particles in weathered mineralization colored with red, orange, and yellow stains. You can pan this material, so a bucket may come in handy. It is messy stuff, however. The Great Excelsior Mine hike is easy for all ages, although it can be wet with all the foliage. This is one of those mines you do not want to enter—it has fumes from the decomposing sulfides, and the support timbers are failing. So just bring a big hammer and break up the tailings. See if you can spot pyrite and quartz crystals as well as microscopic gold. There were a lot of adits and prospects in the hills here, especially up at Goat Mountain, and you can explore for quite some time if you are willing to hike.

10 Olympic Peninsula

See map on page 7.
Land type: Creek
County: Clallam, Jefferson, Grays Harbor
GPS:
A - Shi Shi Beach: 48.255252, -124.684126
B - Ozette: 48.208950, -124.692980 (est.)
C - Yellow Banks: 48.094220, -124.687420
D - Starbuck Mine: 48.014399, -124.677517 (est.)
E - Ruby Beach: 47.709469, -124.414057
Best season: Any; low tide
Land manager: Multiple: Makah Indian Reservation (north), Olympic National Park/Olympic National Forest, Quinault Indian Reservation (Moclips)
Material: Fine gold, black sands, including rare earth and platinum group metals (PGMs)
Tools: Pan, gold wheel, Gold Cube
Vehicle: Any
Special attraction: The Olympics
Accommodations: Hobuck Beach Resort; Ozette Lake CG, motels at Forks and Rialto Beach; Lake Crescent Lodge and Lake Quinault Lodge
Finding the sites: Start with the northernmost locale at Shi Shi Beach. From Third Avenue in Neah Bay, drive west to Cape Flattery Road and turn left (south). After 2.4 miles turn left onto Hobuck Road to cross the bridge, and then turn right onto Makah Passage. Go 1.8 miles, then turn right onto Tsoo-Yess (pronounced and sometimes spelled as Sooes) Beach Road and travel 2 miles to the well-marked trailhead. Take the trail about 3.4 miles (6.8 miles round-trip). Reach the Ozette Beach Placer by going another 4 miles south from the Shi Shi Beach deposit or come in from Lake Ozette. To reach the campground at Lake Ozette, head for Sekiu, where Hoko/Ozette Road and WA 112 intersect. There is no shortcut to Lake Ozette, and there are no services there. From Forks drive 12.3 miles north on US 101, then turn onto WA 113 and drive 10 miles to WA 112, and go another 10.5 miles to the intersection where you could hook up to Neah Bay. Instead, drive 21.2 miles on Hoko Ozette Road to Ozette and go north to the Ozette Placer area, or find Sand Point Trail for Yellow Banks. The Starbuck Mine at the mouth of Cedar Creek is about 7 miles (one-way) south of Yellow Banks. You can also reach it via a

Many of the beaches along the west side of the Olympic Peninsula have concentrated black sands over the years.

7-mile hike north from La Push. To reach La Push from Forks, drive 1.5 miles north on US 101, then turn left onto WA 110. Drive about 7.8 miles, then bear right onto Mora Road and go another 5.1 miles to the beach. Ruby Beach is about 27 miles south of Forks on US 101.

Prospecting

You have to really, really want to go to Shi Shi Beach; there are multiple permits involved. You are on national park land, so you need that permit, and the Makah Indian Reservation also charges a recreation fee. If you want to camp, you need a permit, and if you want to park your car, you have to arrange that with locals—you cannot leave your vehicle at the trailhead. Still, Shi Shi Beach is spectacular on many days of the year, and if the sands are right, you should locate black sands in the biggest creek that drains across the beach. The original mine was up in the terrace, but you cannot dig there. What I did was empty a plastic water bottle and scrape black sand deposits with a nonmagnetic plastic card from my wallet. I came out with a few pounds of concentrates and panned that back home in a tub with dispersant to keep the flour gold from floating away. Ozette is a similar situation—look for the creeks and see if they are washing away the light material and concentrating black sands. My big mistake at Yellow Banks was seeing so much black sand to take home that I poured some of my water out to make more room and regretted it about an hour later. That was a long hike. There are several old placers at Yellow Banks, making it probably the best place to go if you can only go to one place. The Little Wink, Morrow, Yellow Banks, and Main and Bartnes Placer deposits line up where fresh water enters the ocean. Again, do not plan to work on the banks or up on the benches, even though that is where the old-timers were. At the Starbuck Mine, sometimes called Cedar Creek or confused with nearby Sunset Creek and/or Johnson Point, you can only hike in. The logical way is from the north side of the Quillayute River, as there is no bridge and you would have to swim to ford it. The temptation is to go to Rialto Beach in La Push, but that is not where the hike should start. Finally, try Ruby Beach and see if you can spot garnets in the creek. I have never taken much more than a baggie of sand from here, as it is a heavily used area, but if you find a spot where the waves and the creek are concentrating heavies, you should have a representative sample for your collection. Try carefully scraping off the thin black top layer with a straightedge.

CENTRAL WASHINGTON

11 Icicle Creek

Land type: Creek
County: Chelan
GPS:
A - 8-Mile CG: 47.55027, -120.76818
B - Bridge Creek CG: 47.57396, -120.79264
C - Chatter Creek CG: 47.60841, -120.89491
D - Black Pine CG: 47.61025, -120.94521

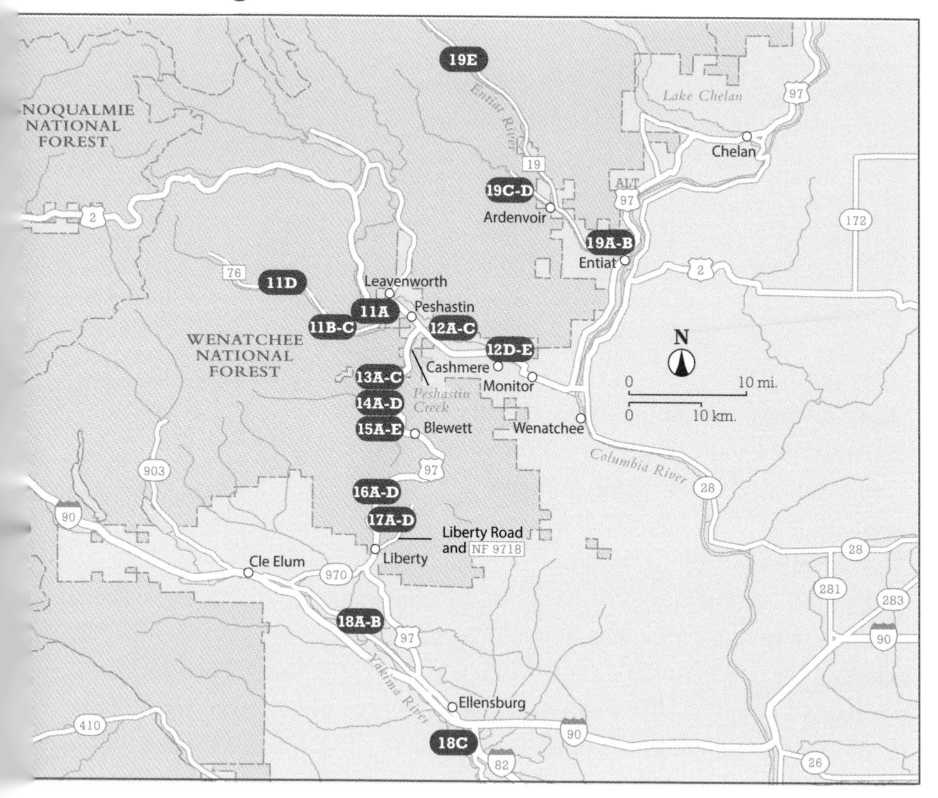

The cliffs along Icicle Creek show considerable mineralization, including veins of white albite feldspar.

Best season: Late summer

Land manager: Okanogan-Wenatchee National Forest

Material: Fine gold, small flakes, black sands

Tools: Pan, sluice, highbanker, dredge

Vehicle: Any

Special attraction: Leavenworth

Accommodations: Blackpine, Rock Island, 8-Mile, and Bridge Creek CGs; dispersed camping all along Icicle Creek; motels in Leavenworth

Finding the sites: Leavenworth is on US 2. To reach Icicle Creek, drive 0.8 mile west on US 2 from the Leavenworth city center and turn left (south) onto Icicle Road. After 6.9 miles turn left onto NF 112 and drive only about 0.3 mile to 8-Mile

Campground. The next spot up the creek is past the Bridge Creek Campground, about 9.1 miles up Icicle Creek Road, or 2.2 miles beyond the turn to 8-Mile. The Chatter Creek site, known as the Addie No. 1 Placer, is 15.5 miles up Icicle Creek Road/NF 7600, or 6.4 miles beyond the Bridge Creek locale. Finally, the Black Pine site is 16.5 miles up Icicle Creek Road/NF 7600; turn left onto NF 7600 and head for the end of the road, about 1.5 miles.

Prospecting

This area is noted for good fine gold, with plenty of black sands and occasional garnets. As you drive up Icicle Creek, look for notable veins of white feldspar in the cliffs, which indicate the interesting geology of the area and contain rose quartz, tourmaline, and other minerals. The 8-Mile Campground at Site A is near the old Bubbly Ranch Mine, and there is a nice bend in the creek below the campground, with rapids cascading through bedrock and boulders. Garnets are plentiful here. Just past Bridge Creek Campground, which also offers good access, there is a popular parking spot looping around a prominent knob. As you get farther up Icicle Creek, the size and amount of the gold persists. Above Chatter Creek Campground at Site C is the site of the old Addie No. 1 Placer, and there are some sizable beaches, bedrock, and boulders to work with. Park at the loop, then make your way down to the creek and work your way upstream, or park closer to the rapids at the head of the island. You will want low water to ford the creek, if possible. Finally, Site D at the Black Pine Campground is near the end of the road near the campground. Bob Jackson (1980) advises to work above Black Pine for the best gold. You can use it as a camp spot and mount hiking expeditions farther up the creek. The Elvira Gold Mine is reported at 47.6352, -120.96781, and a pack trail continues from the end of the road.

12 Wenatchee River

See map on page 29.

Land type: Riverbank

County: Chelan

GPS:

A - Peshastin Creek: 47.557469, -120.575058

B - Wednesday Placer: 47.54397, -120.57262

C - Depot Road: 47.54057, -120.54772

D - Cashmere: 47.52039, -120.45843

E - Overlook: 47.49592, -120.42145

Best season: Late summer

Land manager: Washington Dept. of Fish and Wildlife

Material: Fine gold, small flakes, black sands

Tools: Pan, sluice, highbanker, dredge

Vehicle: Any

Special attraction: Leavenworth

Accommodations: Wenatchee River County Park, Wenatchee Confluence State Park; motels in Leavenworth and Wenatchee; dispersed camping near Blewett

Finding the sites: The site at the mouth of Peshastin Creek, where it enters the Wenatchee, is pretty obvious. From Leavenworth drive east on US 2, past the junction with US 97, about 5.3 miles, and turn left onto Foster Road. Take the immediate left onto Saunders Road, and then drive about 0.3 mile to a right turn to the public fishing access area. Go about 0.3 mile to the large parking area and make your way north along easy trails to the mouth of Peshastin Creek. The Wednesday Placer site is another mile east on US 2; turn on Alice Avenue and circle to Josephine Avenue, about 0.2 mile. To reach the Depot Road site, drive farther east on US 2 to Dryden, about 1.2 miles farther from the Peshastin site. Turn left onto Dryden Avenue, go about 0.2 mile, then turn right onto Main Street. Take the first left onto Depot Road, cross the railroad tracks, and continue 0.7 mile to the turnaround. To reach the Cashmere spot at Site D, go 5.4 miles east from Dryden and turn right onto Cotlets Way, stay left onto Cottage Avenue and cross the bridge, then turn left onto Riverfront Drive. There is parking all along this stretch. The Overlook spot at Site E is on US 2 at a large parking area on the bluff; drive from Cotlets Avenue in Cashmere on US 2 about 2.4 miles.

The Wenatchee River receives gold from Icicle Creek and Peshastin Creek, and during late summer there are multiple access points. This spot is at the mouth of Peshastin Creek.

Prospecting

The Wenatchee contains fine gold with occasional flakes, and there are multiple access points from Leavenworth to the river's mouth at Sunnyside. The Peshastin mouth above Dryden Dam is an excellent spot, with room to roam on a day-use basis. There are multiple public fishing access points from here to Sunnyside—go to https://wdfw.wa.gov/lands/water_access/waterbody/Wenatchee+River/ for all of them. The Wednesday Placer spot at Site B offers more public access to the river and is at a sharp horseshoe bend in the river with good boulders and many rocky zones. The Depot Road locale at Site C is at the Lower Dryden public access and is on another nice inside bend. At Site D at Cashmere, there is an enormous beach on a bend popular with locals; if this is packed, try the nearby Riverside Park off Aplets Way. The Overlook spot at Site E gives a nice view of the river, but it's a steep hike to the water, and you'll need to make your way to the base across from where the islands begin. If it's shallow enough, you can cross to the top end of the big island. You will want to check the river regulations before using bigger equipment on the river and restrict operations to a small window from July 1 to July 31.

13 Ruby Creek

See map on page 29.
Land type: Creek
County: Chelan
GPS:
A - Ingalls Creek: 47.46337, -120.66042
B - Waterfall: 47.44311, -120.66357
C - Ruby Creek: 47.44955, -120.65354
Best season: Late summer
Land manager: Wenatchee National Forest
Material: Fine gold, small flakes, black sands; platinum; garnets on Ruby Creek
Tools: Pan, sluice, highbanker, dredge
Vehicle: Any
Special attraction: Blewett
Accommodations: Dispersed camping on Ruby Creek and Peshastin Creek; seasonal campgrounds at Mineral Springs and Swauk
Finding the sites: The mouth of Ingalls Creek at Site A is right on US 97, about 7.7 miles south from where US 2 and US 97 intersect. The waterfall at Site B is about a quarter-mile hike up Negro Creek from the highway, down the banks of Peshastin Creek and back up to the road. Then follow the trail on the south side of the creek. Ruby Creek is about 0.6 mile north of Negro Creek on the right.

Prospecting

Ingalls Creek is the northern limit of the Blewett District, and you should get a few colors from it. There was once a major placer at the mouth of Ingalls Creek, and the old lost gold mine of Captain Ingalls still entices explorers in the upper reaches of the drainage. Site A is the big parking area near where Ingalls Creek and Peshastin Creek meet, and as of 2022, it has no claims. Ingalls Creek was busy once, but hasn't had a claim in many years. Negro Creek saw considerable activity in the late 1800s, and there are more mines and mineralized areas farther up the creek from Site B at the waterfall. The mouth of Negro Creek is claimed. There are multiple additional claims along Peshastin Creek, so your best bet to get free of mineral claims is to hike up past the falls. There is a decent trail leading from the south end of the old

The narrow gorge carved by Negro Creek contains multiple bends and pools and shows good colors thanks to the many mines and adits above here.

bridge foundation; on the north side, there is a crumbling gated road. The trail bobs and weaves along the creek, eventually leading to a crossing above the falls. Prospecting here requires low water so you can hop across the creek at will. If you continue on the 4WD road above the creek, it eventually doubles back up the hill, and the trail continues along the water. Check the USGS topo map for many more adits, prospects, and cuts, and consult *Discovering Washington's Historic Mines, Volume 2* (2002) for this area.

The old upper Negro Creek Placers are about 5 miles up the creek, and reportedly accessible by road via 12 miles of jeep trail from Shaser Creek. There are multiple old workings up there, including for chromite, and platinum shows up in the black sands. At nearby Ruby Creek there are valid claim markers at the mouth where it is easy to access the water; the road stays maddeningly above the creek as it climbs, but it does loop across the creek several miles from the highway.

14 Blewett

See map on page 29.
Land type: Creek, mine
County: Kittitas
GPS:
A - Historical marker: 47.42355, -120.65909
B - Old Blewett Stamp Mill: 47.42492, -120.65976
C - Matwick Mine: 47.42479, -120.66046
D - Peshastin Creek: 47.42431, -120.65884
Best season: Late summer
Land manager: Wenatchee National Forest
Material: Fine gold, small flakes, black sands
Tools: Pan, hammer
Vehicle: Any
Special attraction: Blewett Mill
Accommodations: Dispersed camping along Peshastin Creek; Swauk CG
Finding the sites: These spots are close together along US 97, about 10.9 miles south from the junction where US 2 and US 97 connect, or 21.2 miles north from where Liberty Road connects to US 97.

Prospecting

Watch for two signs along the highway that compose the historical marker—pull off the road safely to read them. Across the highway at that point, a gravel road leads to the old Blewett Mill to the right. You can take the gravel road if you have a four-wheel-drive vehicle, but if you have a minivan or sedan, you should park off the highway, cross it carefully, and walk the gravel road to the mill. Bring a hammer to break up any interesting white quartz vein material, especially if it has rusty staining. If you feel adventurous, scramble up the hill from the crumbling mill and locate the road; there is an old adit here that is safe to venture into for a few feet. The Northwest Underground team reported that they have seen many better tunnels in the area, and if you enjoy this kind of work, you owe it to yourself to contact them about membership. Most of the land above here is private and posted, so even though you can see from a topo map that there is plenty of mineralization up Culver Creek, you

The old Blewett Mill, just off US 97, dates to around 1906. Look up the hill for one of the many mines in the area.

need permission to explore here. Back down near the historical marker, you can easily access Peshastin Creek, or try a very short drive north to a camping area. This is a club claim, although usually poorly marked. You can try a pan here just to see if it is a club worth joining; my hunch is that you will like what you see. If there is anyone working the claim, chat them up and get more up-to-date information. The remains of an old arrastra—a water-powered stone grinding apparatus, is behind the guardrail at 47.4225, -120.6593.

For historical research, Washington Geological Survey Bulletin No. 6, "Geology and Ore Deposits of the Blewett Mining District," by Charles E. Weaver (1911), is a good start, and it is available in PDF format. Northwest Underground Explorations also has great (and more updated) information in *Discovering Washington's Historic Mines, Volume 2* (2002).

15 Peshastin Creek

See map on page 29.
Land type: Creek
County: Kittitas
GPS:
A - Wilson: 47.41244, -120.65883
B - Caledonia: 47.42002, -120.65762
C - West side: 47.43059, -120.65629
D - Cook: 47.43107, -120.65481
E - Solita: 47.43268, -120.65735
Best season: Late summer
Land manager: Wenatchee National Forest
Material: Fine gold, small flakes, black sands
Tools: Pan, sluice, highbanker, dredge
Vehicle: Any
Special attraction: Blewett
Accommodations: Swauk CG; dispersed camping at multiple sites, but they're right on the highway and noisy.
Finding the sites: The Peshastin Creek drainage contains good gold, so we've already listed three spots—at the mouth, under the Wenatchee River; above the mouth of Negro Creek; and at Blewett. The west side, Cook, and Solita Placers are north of Blewett, about 0.7 mile. The Caledonia locale is about 0.3 mile south of Blewett, and the Wilson site is about 0.9 mile south of the Blewett historical marker.

Prospecting

Peshastin Creek runs briskly through here, often straight along the highway, cutting underneath in some interesting culverts that leave you wondering why they cannot put Hungarian riffles inside culvert pipe. The gold is excellent, although fine, and there are good black sands. The only caution here is to check for active claim markers, mostly from clubs; you can always grab a quick sample to see if the spots are good enough to join, but you would not be able to set up equipment without joining. Each of these pull-outs between Negro Creek and the turnoff for Old Blewett Pass via NF 7320 is worth checking

Peshastin Creek features ample bedrock, large boulders, and excellent gravels to work for good gold, especially near Blewett.

for a camp spot or open access. The Wilson Placer spot is on a nice bend in the creek, with ample bedrock to poke around on, especially near the bridge. There are some good holes, probably from past dredgers, at the Caledonia site. Look for prospects on the east side of the road near the parking spot. North of Blewett, the Cook and Solita Placers are both on excellent bends in the creek, and there is access through the guardrails to nice camping areas on the water. If the price of gold drops significantly, many of these claims may free up, so for now these sites remain in this guide, but use caution.

16 Swauk Creek

See map on page 29.
Land type: Creek
County: Kittitas
GPS:
A - Swauk CG: 47.327934, -120.661938
B - Durst Creek Road: 47.32355, -120.67688
C - NF 9705: 47.30436, -120.69554
D - Bovee: 47.29587, -120.69995
Best season: Late summer
Land manager: Wenatchee National Forest, BLM–Wenatchee
Material: Fine gold, small flakes, black sands
Tools: Pan, sluice, highbanker, dredge
Vehicle: Any
Special attraction: Liberty
Accommodations: Dispersed camping on both sides of US 97. Developed campgrounds at Mineral and Swauk are seasonal; Williams Creek CG is open more often.
Finding the sites: The 7-mile Swauk Creek run is open for prospecting from Liberty Road north to the Swauk Campground, with the exception of a couple spots that are private and have mailboxes on their driveways. Other than that, you are free to roam. Swauk Campground is 25.1 miles south of the intersection of US 97 and US 2, or 10.1 miles north of the intersection of US 97 and WA 970.

Prospecting

Swauk Creek is "withdrawn from mineral entry" here, or closed to claims. That means it is open to the public, so you just need good access. Camping is iffy along here, but only because it gets noisy due to truck traffic on US 97, and you are rarely more than 300 yards from the centerline. Starting from the north, the Swauk Campground site is about as far north as you want to be in the good mineralized area. The creek is quite small here; you are close to its headwaters. Durst Creek Road not only provides good parking and creek access, but there are many mines along the road as you go up. Northwest Underground Explorations covered this area in *Discovering Washington's*

Nice chunky gold from Swauk Creek

Historic Mines, Volume 2 (2002). The next coordinates, on NF 9705, used to provide access to great creek spots on a dirt road that paralleled US 97 south. Unfortunately, boulders now block it, but it's an easy hike. I have not searched for a ford at the Bovee Placer site that might lead to that road, but you never know. I did not list coordinates for the Mineral Springs Resort, but it would be a good place to stop in if you visit between Memorial Day and Labor Day, and Mineral Springs Campground is a good spot to stage from. Lower down, there is a giant placer operation at Cedar Creek, on the east side of the highway, but it is closed to the public.

17 Liberty

See map on page 29.

Land type: Creek

County: Kittitas

GPS:

A - Liberty town: 47.25371, -120.66686

B - Williams Creek: 47.24544, -120.68477

C - Cougar Gulch: 47.28309, -120.64871

D - Liberty Mine: 47.29678, -120.64187

Best season: Late summer

Land manager: Wenatchee National Forest

Material: Fine gold, small flakes, possible wire gold

Tools: Pan, sluice, highbanker, dredge

Vehicle: Any; 4WD suggested for exploring

Special attraction: Liberty Mine

Accommodations: Williams Creek CG; dispersed camping up Pine Gulch, Lion Gulch, and Cougar Gulch

Finding the sites: The Liberty Road turn on US 97 is about 3 miles north of the junction with WA 970 and US 97. Turn here and drive 0.7 mile, then turn right and circle down to the Williams Creek Campground. There is camping on both sides of the creek; this is a BLM site. The town of Liberty is 1.7 miles from US 97. To reach the site on Cougar Gulch, continue past Liberty 0.8 mile, then take the left onto NF 9718 and drive 1.8 miles. To reach Lion Gulch and the Liberty Mine from Williams Creek Campground, again drive toward Liberty, but do not go into town. Instead, turn left after about 0.7 mile from the turn to the campground and drive up Cougar Gulch Road/NF 9712. About 4.2 miles up, look for a right turn going up the hill, and wind around less than 0.1 mile to a flat road coming in from your right. It is an easy hike to the mine entrance.

Prospecting

There are hundreds of interesting places in the hills and creeks around Liberty, so this is just a starter list. Liberty is famous for its wire gold, especially from the Ace of Diamonds Mine, where delicate gold crystals reside in calcite, just waiting for muriatic acid to etch away the host and reveal beautiful strings,

This awesome dual-engine setup on Williams Creek is efficiently operated by a group of women.

lines, wires, and other shapes. The Williams Creek Campground is a great central location for a base camp, and if you are in the Prospectors Plus Club, you can dredge here during the season, with the right permits and so forth. On the side of the creek closest to the highway, you can find one of the many mine shafts driven into cemented gravels, searching for pay dirt along bedrock in ancient creek channels. There are dozens of such shafts in this area, but most are private and dangerous, being rarely timbered or braced.

You owe it to yourself to stop in at Liberty and check out the replica of an old arrastra, a crude stone wheel that could grind gold ore into powder to loosen free gold. Liberty used to have services, but by 2022, even the little store was shut down, which is a shame. There was a time when you could get good local info from the store owner if you bought a beverage or stocked up on ice. He was working a deep mine sunk through the gravels to bedrock.

Farther up from Liberty on Cougar Gulch, there are plenty of reminders of past activity. The coordinates here are at Billy Goat Gulch, just before the hairpin turn. You should find an old adit right by the main road, possibly the Cascade Chief. Farther up this little road are the remains of an old bunkhouse to the right. You can sample anywhere you can find access to water that is not posted with an active claim. The creek at the Liberty Mine is claimed by the Prospectors Plus Club, and the last time we were there, a team was working the creek near the culvert. The road is a true jeep trail going up, but it's not far to the turn to the mine. As usual, enter at your own risk—there are cave-ins not far from the entrance.

Finally, keep an eye out for Rob Repin's PayDirt, the third driveway north on US 97 from milepost 152. Rob is super knowledgeable about the Liberty area and owns multiple active claims, both lode and placer. He has some YouTube videos you can check out too. During the summer, from Memorial Day to Labor Day, Rob is usually open on weekends from noon to 4 p.m., selling scoops of pay dirt from his mines. It's not guaranteed, and he may opt out at any time from selling his pay dirt. If he's open, he'll let you gaze into his microscope and check out the crystals he has collected that look like giraffes and other forms, and if you buy a scoop of his pay dirt, he will let you pan it in his big tub. Rob has a clever, science-based approach to panning—he never swirls his pan, but instead sloshes it from side to side, focused on getting the gold and heavies to pile up behind the very first riffle in his pan. Then he can tilt the pan almost straight up and drain off the unwanted material. The heavies are stuck on that first riffle, and he can clean a pan in no time.

18 Yakima River

See map on page 29.
Land type: Riverbank
County: Kittitas
GPS:
A - Thorp: 47.10088, -120.70155
B - Magpie Canyon: 47.09636, -120.68914
C - Ringer: 46.92658, -120.51665
Best season: Late summer
Land manager: Washington Dept. of Fish and Wildlife
Material: Fine gold, small flakes, black sands
Tools: Pan, sluice, highbanker, dredge
Vehicle: Any
Special attraction: Roslyn
Accommodations: Ellensburg KOA, Williams Creek CG (see Liberty); no dispersed camping along the Yakima; try USFS land toward Frost Mountain or up the Teanaway River near Cle Elum.
Finding the sites: To reach the Thorp locale at Site A, start at the intersection of WA 10 and WA 970 at Teanaway and drive east on WA 10 for 9 miles, then take a right onto Thorp Highway. Drive about 0.8 mile to the bridge, cross it, and look for an immediate left to the public access area. In 2022 the bridge was under repair, but it should eventually be back in service. To reach Magpie Canyon, return to WA 10 and turn right, then drive 1.5 miles and find a safe place to park. You will have to backtrack about 0.1 mile to reach Magpie Canyon. To reach Site C at Ringer, get back on I-90, head east to I-82 and go south, then take exit 3 onto WA 821 and drive 0.4 mile to Canyon Road. Go right and take the first left onto Ringer Loop Road, go 0.3 mile, and take a left down to the water.

Prospecting

The Yakima is only open for dredging in the month of August, according to the "Gold and Fish" pamphlet, so be prepared to only pan here. These locales are sporadically productive, with good colors, as long as the river is low enough. If you could mount an expedition via watercraft to the mouth of Swauk Creek at 47.1237, -120.7379, there is a long, productive gravel bar

Collect multiple samples in the field and process carefully back in camp.

downstream from the Swauk for a good distance. It might be interesting to check out the Teanaway River mouth, just for completeness. Don't be afraid to rock-hound here either, as the famed Ellensburg Blue agate is possible in these gravels. I haven't checked for access from the other side of the river; the Palouse to Cascades State Park Trail might be a good option. The access at Site A near the Thorp bridge is good, but there is only exposed gravel showing late in the summer. At Site B at Magpie Canyon there is easy river access, and for the next couple of miles, before the railroad tracks come back in the picture, there is more access, such as at the Thorp takeout. One final spot that I haven't visited yet is the Perry Placer near Castle Rock, reachable only via boat or from the Iron Horse Trail. It is somewhere near 47.1137, -120.7186, where the river makes a big bend around the basalt outcrop.

19 Entiat

See map on page 29.
Land type: Creek, riverbank
County: Chelan
GPS:
A - Mouth: 47.71444, -120.27398
B - Rex Mine: 47.71474, -120.27352
C - Indian Creek: 47.74193, -120.46148
D - Pine Flats CG: 47.75873, -120.42404
E - Box Canyon: 47.93795, -120.52178
Best season: Late summer
Land manager: Wenatchee National Forest
Material: Fine gold, small flakes, black sands; platinum
Tools: Pan, sluice, highbanker, dredge, hammer
Vehicle: 4WD required for Crum Canyon, suggested elsewhere
Special attractions: Upper Entiat Valley, Entiat Falls
Accommodations: Daroga State Park, Pine Flats CG, multiple campgrounds in the upper Entiat; dispersed camping along both rivers and at Indian Creek; motels at Chelan and Wenatchee
Finding the sites: About 0.6 mile west of US 97-Alt on Entiat River Road, look for a pull-out and a gravel road leading to the water. This is a local fishing hole, and there is room to roam here. To reach Site B and the Rex Mine, drive another 2 miles to the turn on the right to Crum Canyon Road/CR 301; go 2.1 miles, staying right to avoid Osbourne Road and to stay on Crum Canyon Road. Go another 1.1 miles and look for decaying structures. To reach Site C at Indian Creek, return to Entiat River Road and continue to Ardenvoir, about 9.8 miles total up Entiat River Road, then turn left onto Mad River Road. Take it for 2 miles, then turn left onto NF 5800. Drive 2.3 miles, then turn right onto NF 5808 and drive 1.7 miles to the hairpin turn on Indian Creek. From Indian Creek you apparently can reach Entiat Summit Road, which leads to old lode mines reported on Maverick Peak. To find Pine Flats Campground, return to the junction of Mad River Road and NF 5800, and turn left onto NF 5700/Mad River Road. The left turn down to the campground is at about 1.8 miles, before you start climbing up the hill. To reach Box Canyon on the Entiat River, return to Ardenvoir and drive upstream about 18 miles.

Old ruins near Ardenvoir, where Indian Creek joins the Mad River

Prospecting

The Entiat and Mad Rivers area contains platinum in the black sands, so hang onto your concentrates. Geologists never divided the district, so it covers almost 800 square miles. You can pan at Site A, as this was the locale of the Entiat River Placers, but the dams on the Columbia River have flooded the actual mouth of the Entiat. The mines on Crum Canyon date to the 1930s and produced gold and silver in vein systems. Bring a hammer to the Rex Mine, as it produced more than $170,000 by 1930, exploiting gold and silver in two oxidized quartz veins, 3 to 12 inches in width, in decomposed gneiss of the Chelan Complex. You can still see the two-stamp mill along the road near the coordinates. The Sunshine, Savage, and Pangborn Mines reportedly are also up in these desolate hills south of the Rex Mine if you feel like doing more research and exploring (start at Mindat.org to investigate the Entiat District). There isn't a lot of water in Indian Creek late in the season, and you need to keep an eye out for active claim markers, as one of the clubs erected signs up here one year. Still, Indian Creek is worth investigating, and the section downstream starting at 47.7384, -120.4508 was unclaimed as of 2022. I did better on the Mad River; I found open spots that were accessible all the way from the ruins at the intersection of NF 5800 and Mad River Road to Pine Flats. There's good color at Pine Flats Campground; be sure to work your way up the river (it's the size of a creek, actually) because there is a nice canyon with good bedrock traps there. Back on the Entiat River, you can pan your way from about 7 miles above Ardenvoir all the way to Box Canyon at Site E and sample as you go. The dredging season is currently restricted to a two-week period on the Entiat River to the falls, from July 16 to July 31. The area above the falls is open much longer.

20 Lake Chelan

Land type: Riverbank
County: Chelan
GPS:
A - Stehekin: 48.37034, -120.75458
B - Lucerne: 48.20123, -120.59493
Best season: Late summer
Land manager: Lake Chelan National Recreation Area
Material: Fine gold, small flakes, black sands
Tools: Pan, trowel, snuffer bottle
Vehicle: None; ferry only

Central Washington North

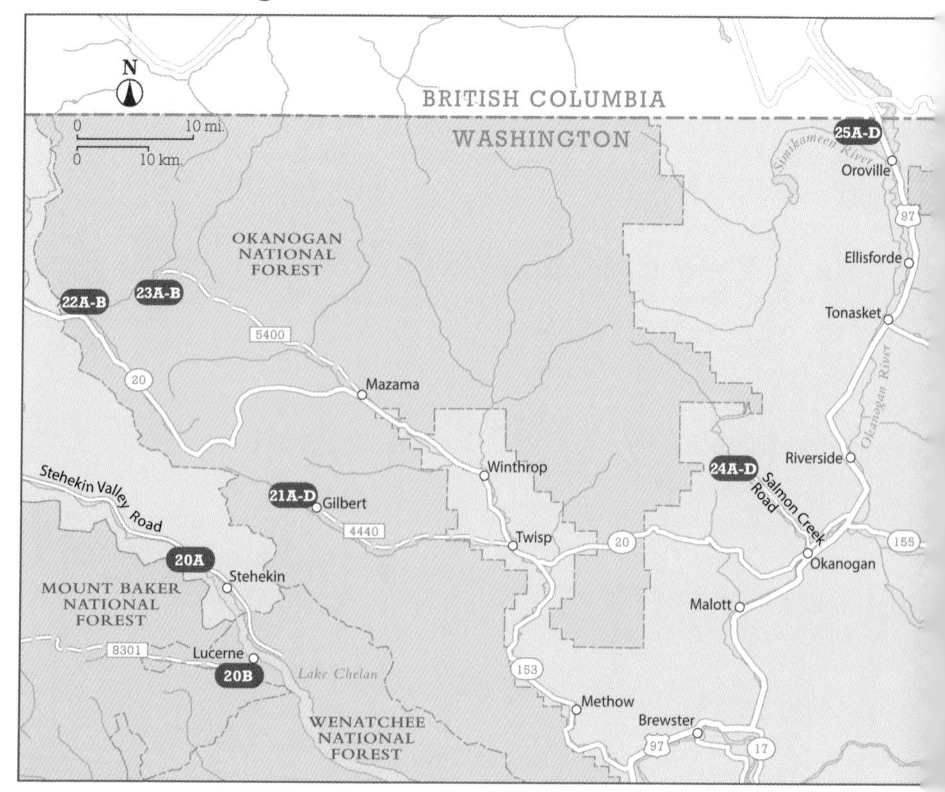

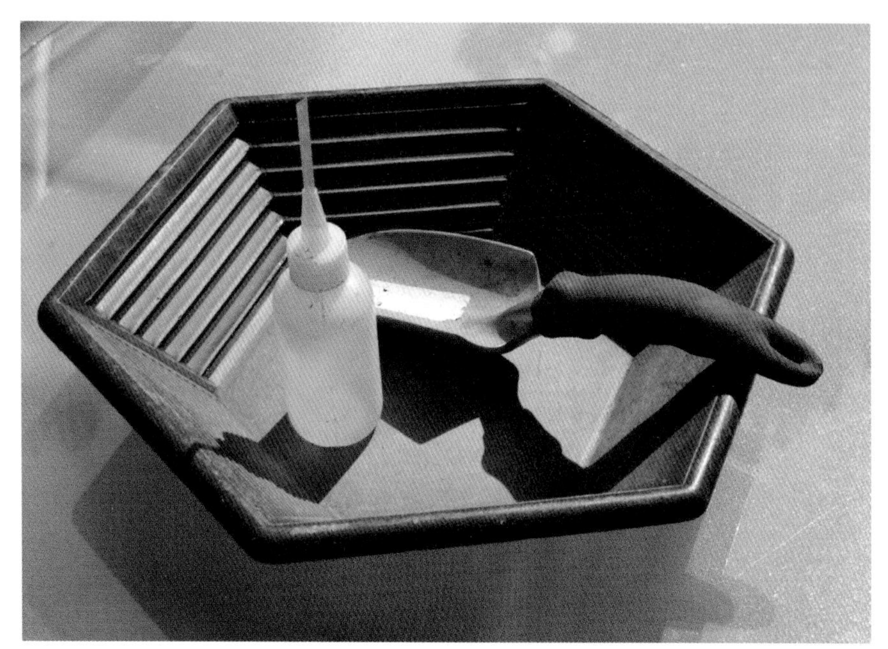

You will have to pack light for the boat trip and bicycle jaunt to Stehekin, at the upper end of Lake Chelan.

Special attraction: Rainbow Falls

Accommodations: North Cascades Lodge; Purple Point CG at Stehekin; more camping along Stehekin River (dispersed camping requires permit in national park)

Finding the sites: This is one of the most challenging sites to access in the book because you cannot simply drive there. Instead, you need to take the *Lady of the Lake* ferry from Chelan to Stehekin. Of course, some of you may choose Chelan Seaplanes to fly up there, and others may bring their own boats. However, the ferry is the simplest way to go; consult http://ladyofthelake.com for the schedule. The ferry makes a quick stop at Lucerne, and you can stop there, check it out, and take the next ferry that comes through to Stehekin. Once at the upper end of the lake, our next step was to rent bicycles at Stehekin Discovery Bikes (http://stehekindiscoverybikes.com). We then pedaled to Stehekin Pastry Co. (http://stehekinpastry.com) and ordered up some lunches for the ride. From there we checked out Rainbow Falls. Finally, we pedaled to the panning site, about 7 miles total from the dock, all uphill, but it was great zipping back down.

Prospecting

It would be a lot of fun to use a boat on the lake and journey all over on your own schedule. The Holden Mine on Railroad Creek, above Lucerne, was a major producer, starting in 1887. Although copper was the main commodity, the Holden produced gold, zinc, silver, molybdenum, nickel, and lead as by-products. It is now a Lutheran retreat and summer camp and the site of a major reclamation project. There is evidence of the old Railroad Creek Placer at 48.1898, -120.6115. The trails above Holden access lots of interesting geology in the Glacier Peak Wilderness. Just across the lake from Lucerne at Meadow Creek, there are multiple prospects in the steep hills. From the landing at Stehekin, there are pack trails leading to the high peaks at Cascade Pass, Horseshoe Basin, Park Creek Pass, and elsewhere. There's even a placer mine where swimmers now enjoy Lakeside Park, just west of Chelan. If all you do is get some quick samples of black sands and color from the mouth of Railroad Creek at Lucerne, or skip that and go on up to Stehekin for a few pans, you're still the better for it.

21 Gilbert

See map on page 50.
Land type: Creek
County: Okanogan
GPS:
A - Payday: 48.46028, -120.56000
B - Turnaround: 48.45844, -120.56769
C - Gilbert Cabin: 48.45732, -120.56227
D - Road's End CG: 48.46144, -120.57828
Best season: Late summer
Land manager: Okanogan-Wenatchee National Forest
Material: Fine gold, small flakes, black sands
Tools: Pan, sluice, highbanker, dredge
Vehicle: Any; 4WD suggested
Special attraction: Gilbert Cabin
Accommodations: Campground at Road's End; dispersed camping throughout area; motels at Twisp
Finding the sites: These are easy sites to find—just drive to the end of the road. From the intersection of WA 20 and Second Avenue in Twisp, turn west onto Second Avenue. It will become Twisp River Road/NF 4440. After about 24 miles you will see the Payday site, where a beautiful stream called North Creek comes down to the road from Gilbert Mountain. Another 0.5 mile west on the main road is a turnaround to the right, and a road headed back to the Gilbert Cabin is on the left. Continue another 0.5 mile to the Road's End Campground, almost exactly 25 miles from town.

Prospecting

There is good color at the Road's End Campground, and you can explore up farther as the Twisp River becomes little more than a boisterous creek. Alternatively, you can follow the Twisp east about 0.8 mile to about 48.4537, -120.5616, where North Creek meets the river. The road curves away from the river here at the end, so adjust accordingly. There are many options here—you could start at the cabin, find North Creek, and follow it down to its mouth. On the south side of the river, headed up Crescent Mountain, you can

This old cabin is the last remaining upright structure from Gilbert's glory days.

locate the old Crescent Mine at 48.4514, -120.5802. If you plan to spend a lot of time up here, consult Northwest Underground Explorations' *Discovering Washington's Historic Mines, Volume 3* (2006) for excellent information about the Gilbert area. Like most rivers in Washington, the dredging season is very restricted on the Twisp River, so plan accordingly.

22 Ruby Arm

See map on page 50.
Land type: Creek
County: Whatcom
GPS:
A - Lazy Tar Heel: 48.70483, -120.94772
B - Johnnie S.: 48.70646, -120.91853
Best season: Late summer
Land manager: Mount Baker–Snoqualmie National Forest
Material: Fine gold, small flakes, black sands; garnets
Tools: Pan, screens, and buckets only
Vehicle: Any
Special attractions: Ross Reservoir, Diablo Reservoir, Beebe Cabin
Accommodations: Newhalem CG; dispersed camping along Chancellor Trail by backpack
Finding the sites: From Newhalem on US 20/North Cascades Highway, drive east for 19.1 miles to the westernmost site, at the old Lazy Tar Heel Placer operation. There is good parking here. About 1.4 miles farther east is the Johnnie S. claim and the parking area for the Chancellor trailhead. This is where Canyon Creek and Granite Creek form up into Ruby Creek. I did not find good color on Granite Creek, so avoid it.

Prospecting

The Ruby Arm of Ross Reservoir contains red garnets (not rubies), and there are plenty in most pans. They are mostly deep red and purple and are easy to spot if you have a loupe, magnifying glass, or hand lens. Screening would be a good idea to pick out the bigger garnets. When Ross Reservoir formed, it flooded out many old placer operations on Ruby Creek. Once you clear the national park land, you can start scouting for good spots. The Lazy Tar Heel locale is on a nice bend and has adequate parking. Even better, this stretch has been withdrawn from mineral entry—nobody can claim it, so it will always be open. Another good bend is farther east, but parking is not as good. There is a small, gravelly beach below the parking area at the Johnnie S., and if you are in a rush, you can pan a few samples there. For a stretch objective, hike

You can reach the Chancellor Trail via the parking area at the Johnnie S. locale and find good colors up Canyon Creek as far as you want to go.

the Chancellor Trail up Canyon Creek. Be on the lookout for claim markers, but you can settle for just about any spot along the water here for the first half mile. The trail is an easy hike, and there is plenty of bedrock to trap heavy material. It's way too far to hike in big equipment, which is just as well. You would need to contact the Forest Service about a dredge permit.

23 Slate Creek

See map on page 50.
Land type: Creek
County: Whatcom
GPS:
A - Bridge: 48.74171, -120.71168
B - Ford: 48.72893, -120.74129
Best season: Late summer
Land manager: Okanogan-Wenatchee National Forest
Material: Fine gold, small flakes, black sands; occasional garnets
Tools: Pan, sluice, highbanker, dredge
Vehicle: 4WD required
Special attraction: Slate Pass
Accommodations: Meadows CG near Harts Pass; dispersed camping up and down Slate Creek; B&Bs in Mazama; motels in Winthrop
Finding the sites: The little town of Mazama is about 19 miles northwest of Winthrop via WA 20. Turn onto Lost River Road and drive 0.2 mile to Mazama, then turn left to stay on Lost River. Drive 6.7 miles and continue, now on NF 050. Next the road becomes NF 5400; stay on that for 2.1 miles, then turn sharply right to stay on NF 5400 headed for Harts Pass and Slate Pass. Drive 9.8 miles to reach Harts Pass, then turn left onto NF 700, which heads down the hill sharply. Drive 2.8 miles to the gated access road NF 700 to Barron. The main road is now NF 374; drive 0.8 mile to the bridge locale. The mouth of Bonita Creek is about 0.3 mile farther downhill. The Grizzly Placer is about 0.5 mile farther down NF 374, and the ford across Slate Creek is another mile farther. Even if you have the right rig and can ford the creek, NF 374 is in poor condition, and you will likely have to walk the additional 4 miles downhill, along Slate Creek, to reach the old mining camp at Chancellor at 48.7581, -120.7947.

Prospecting

There are countless old prospects and mines in this area; Slate Creek was a good producer, but it had wretched winter storms that resulted in avalanches, cave-ins, and other hardships. If you want to explore the hard-rock mines in this region, consult *Discovering Washington's Historic Mines, Volume 3* (2006).

The drive up to Harts Pass provides excellent views of Methow Valley, but your passengers may not appreciate the steep drop.

Note, however, that the dozens of properties at the old town of Barron are on patented claims and now private. The bridge area at Site A is open; Slate Creek has no claims below this point, according to MineCache.com. The gold on Slate Creek is impressive, which is why there are now claims all the way to the ford on Slate Creek. If you could get a dredge down here in the right season and situate it in a decent hole, you might do quite well, but most of us will settle for a pan. I had good success below Bonita Creek, with colors in every pan and many good-size flakes. I still have not made the journey past the ford, given the difficulty in reaching it. In years past I have seen claim markers along Slate Creek, particularly from the Washington Prospectors Mining Association (WPMA). This is one of those areas that would easily make a membership worthwhile. The country is remote, beautiful, and heavily mineralized—an excellent combination.

24 Ruby

See map on page 50.
Land type: Creek, mine tailings
County: King
GPS:
A - Ruby City: 48.49791, -119.72389
B - Last Chance: 48.49157, -119.74141
C - Arlington Mine: 48.47224, -119.73656
D - Fourth of July: 48.47657, -119.72869
Best season: Late summer
Land manager: Okanogan-Wenatchee National Forest
Material: Fine gold, ore samples
Tools: Pan, hammer
Vehicle: 4WD required beyond Ruby
Special attraction: Conconully
Accommodations: Conconully State Park, Rock Lakes CG; dispersed camping at Last Chance and beyond
Finding the sites: Ruby is 11.8 miles northwest of Okanogan via Salmon Creek Road. When you reach the junction with the Ruby Grade Road, you should see a sign. To reach the mines, drive up Ruby Grade 1.2 miles and turn left onto Peacock Mountain Road. Drive 0.4 mile and turn left onto Buzzard Lake Road. After 0.4 mile look for an access road to your right, or continue another 0.2 mile and walk in from above the diggings. This is the Last Chance Mine. There are more old tailings above the road. Continue 1.6 miles up the hill to reach the Arlington Mine, and another 1.2 miles to reach the Fourth of July locale.

Prospecting

Ruby was named not in error for garnets but for a particular variety of silver ore named pyrargyrite, also known as ruby silver. Early prospectors struck a ledge about 18 feet wide that ran uniformly from wall to wall with a value of $14 in gold and silver—but mostly silver. At the Fourth of July Mine, average assays from the vein showed the ore to be worth over $100 per ton. You should be able to pan color from Salmon Creek at Ruby; there is plenty of mineralization around Conconully to serve as a good source, even with the

This opening near the Arlington Mine displayed considerable mineralization, and it did not require a descent into the dark to retrieve a good sample.

dam there. Ruby was quite a town back in the day, but with few concrete foundations to mark it now, you would never guess its history. Still, there are gravels, and even a little shade, along the creek at Site A, and it runs most of the summer. The rest of the coordinates are hard-rock lode mines, rather than panning, but it's always interesting to see the mines that fed a placer area. The many mine dumps and tailings piles up Ruby Grade offer a great opportunity to swing a hammer; we found green malachite and blue azurite, black biotite mica, shiny native silver, and specks of gold at the Last Chance tailings, so we stopped whenever we saw diggings, tailings, workings, or other signs. Each site was rewarding.

25 Similkameen River

See map on page 50.
Land type: Shallow river
County: Okanogan
GPS:
A - Falls: 48.96592, -119.50129
B - Miners Flat Recreation Site: 48.97959, -119.52979
C - Rich Bar: 48.98231, -119.53749
D - Similkameen access: 48.98909, -119.58013
Best season: Late summer
Land manager: BLM, Washington Dept. of Fish & Wildlife
Material: Fine gold, small flakes, black sands; platinum
Tools: Pan, sluice, highbanker, dredge
Vehicle: Any
Special attraction: Nighthawk
Accommodations: Palmer Lake CG, Conconully State Park; dispersed camping along the Similkameen
Finding the sites: The city of Oroville is on US 97, about 5 miles from the Canadian border. Start there and drive west on Loomis-Oroville Highway, along the Similkameen River, for a total of about 3.8 miles. Look for a left turn onto Enloe Dam Road and go down the hill about 0.5 mile to the parking area. This is Site A. The next spot is a dry-camp area about 6.5 miles from the Oroville city center, or another 2.7 miles past the turn for Enloe Dam. Look for turns down to the river. Access to the old Rich Bar Placer operation is about 0.4 mile farther east. The final access point listed here is another 2.2 miles east, where there are camp spots and multiple dirt tracks leading to the river's edge.

Prospecting

The Enloe Dam site is not my favorite—it is noisy and misty until August, and the dam's design freaks me out (the way the water slips over the top). However, there are good exposures of bedrock above it, and it's a decent place to grab a sample. The rest of the sites listed for the Similkameen are just a starter, as there is additional access above Nighthawk, including public access spots run by the Washington Department of Fish & Wildlife. The Miners Flat

Dredgers working a shallow stretch of the Similkameen River. This river is famous for giving up platinum as well as gold.

Recreation Area is sizable and offers many dry-camping sites as well as river access. The Similkameen area at Site D is also a public area maintained by the BLM, with numerous camping spots along the shore. There are multiple prospects dotting the hills on both sides of the river, but you will need a good 4WD vehicle to access them, and you will run into private land. There is a longer dredging season here, if you are so inclined, and your black sands are bound to contain traces of platinum, iridium, and other rare earth elements (REEs) or platinum group metals (PGMs). The source of the platinum is probably in Canada, while the gold is likely more local. The river is quite shallow by late summer, and it is common to see dredges right in the middle of the waterway. Panning works quite well anywhere you can find an inside bend and rocks and boulders. If you check out the area via Google Earth, you will see dozens of prospects and tailings piles, some with roads worth exploring. Consult *Discovering Washington's Historic Mines, Volume 3* (2006) for more information if you want to prospect Ellemeham Mountain or other surrounding areas.

SOUTHERN WASHINGTON

26 Washougal River

Land type: Riverbank, creek
County: Skamania
GPS:
A - Dougan Falls: 45.67278, -122.15426
B - Above swimming hole: 45.70763, -122.12431
C - Prospector Creek: 45.74303, -122.11351
Best season: Late summer
Land manager: Gifford Pinchot National Forest, Yacolt Burn State Forest
Material: Fine gold, small flakes, black sands
Tools: Pan, sluice, highbanker, dredge
Vehicle: Any; 4WD suggested beyond Dougan Falls
Special attraction: Dougan Falls
Accommodations: Dougan Creek CG; dispersed camping along the Washougal above the falls, mouth of Prospector Creek
Finding the sites: From Washougal on WA 14, drive east about 10 miles to Salmon Falls Road and turn left (north). Drive 3.4 miles to Washougal River Road (which you could have picked up in Washougal, but it is a longer route). Turn right and drive 5.7 miles on Washougal River Road to Dougan Falls. To reach one of the access points above the popular swimming areas, turn right after the falls and follow the river for 3.8 miles on Road W-2000. To reach Prospector Creek, drive from the bridge on W-2000 for 6.8 miles, then go right for 1.1 miles, turn left and drive 0.5 mile, and then turn left for 1.4 miles. The roads are now gated beyond here along Prospector Creek.

Prospecting

It isn't easy to get a sample at Dougan Falls, but there are some footpaths down. Prospectors often dream about plunge pools below waterfalls, and this would be one to consider. The local timber company is continually refining access to the area, as Dougan Falls is a popular (read: crowded, noisy, boisterous, and barely under control) swimming spot in the hot summer months, so you need parking permits and planning if you want to sample along here.

Southern Washington

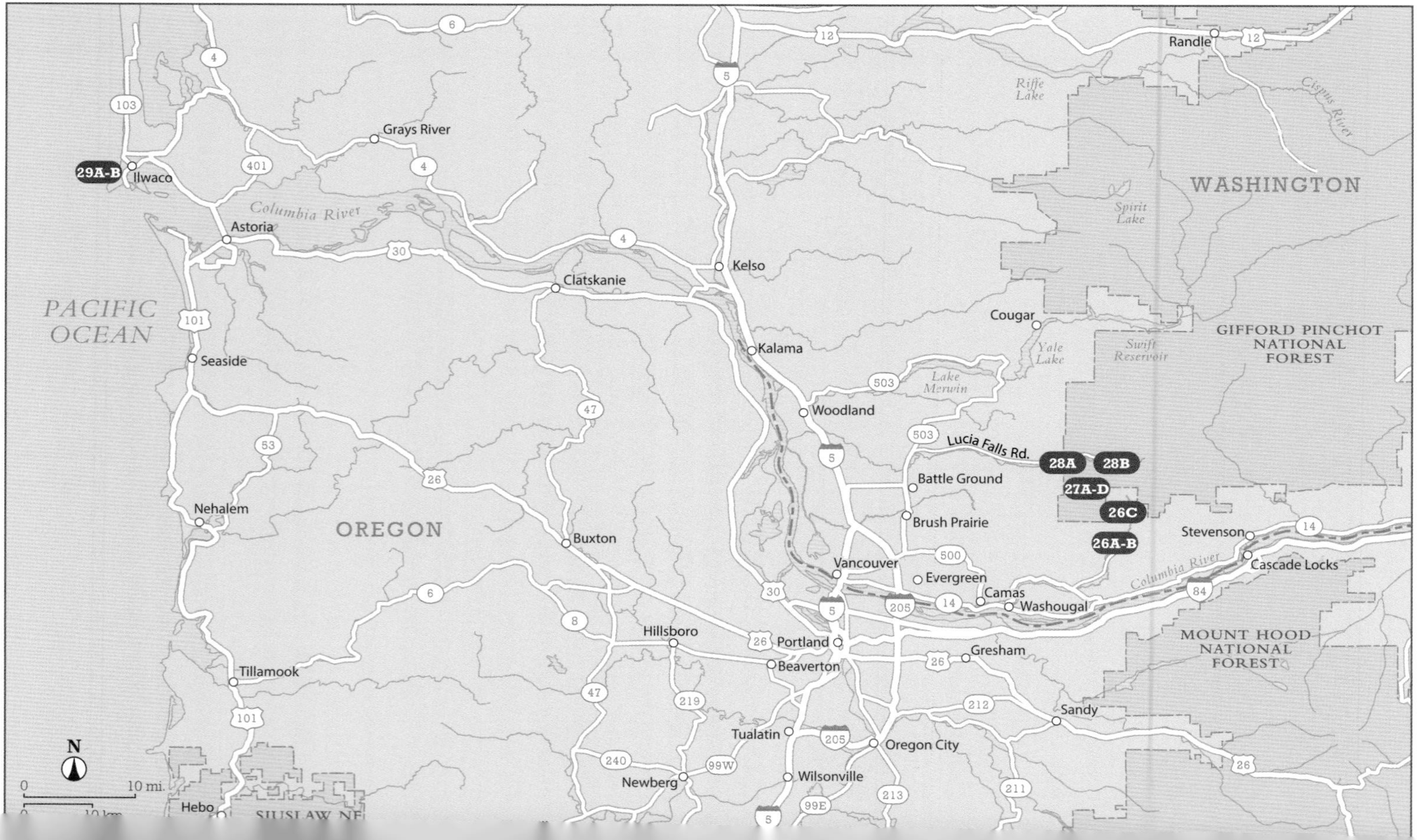

PACIFIC OCEAN
WASHINGTON
OREGON
GIFFORD PINCHOT NATIONAL FOREST
MOUNT HOOD NATIONAL FOREST
SIUSLAW NF
Columbia River
Cispus River
Riffe Lake
Spirit Lake
Swift Reservoir
Yale Lake
Lake Merwin
Lucia Falls Rd.
29A-B
28A
28B
27A-D
26C
26A-B
Randle
Grays River
Ilwaco
Astoria
Kelso
Clatskanie
Cougar
Seaside
Kalama
Woodland
Battle Ground
Brush Prairie
Nehalem
Buxton
Stevenson
Cascade Locks
Vancouver
Evergreen
Camas
Washougal
Hillsboro
Portland
Beaverton
Gresham
Tillamook
Sandy
Tualatin
Oregon City
Wilsonville
Newberg
Hebo
6
4
12
5
103
401
30
101
503
47
53
26
14
500
84
8
205
219
212
240
99W
99E
213
211
N
0
10 mi.

The Washougal River almost overwhelms Dougan Falls during the winter months, but it settles down by summertime.

A better option is to keep going to easier access points once you reach US Forest Service land. Site B has ample bedrock during the low water months. The best gold on the Washougal is far, far up at Prospector Creek (beyond Site C); the road used to reach farther but now is blocked. You'll have to hike to the mouth of Prospector Creek at 45.7404, -122.1293. One good guide to this area is the Washington Department of Natural Resources Information Circular 60, *St. Helens and Washougal Mining Districts of the Southern Cascades of Washington* (1977), by Wayne S. Moen. The copper used to erect a statue to Sacajawea in Portland came from this district. The report is easily available as a PDF using a Google search. The lode mines in this area are notoriously difficult to find and are generally overgrown with lush vegetation, and the samples are not inspiring. Still, the panning is good, and there is a one-month dredging season in August.

27 Copper Creek

See map on page 64.
Land type: Creek
County: Skamania
GPS:
A - GPAA bridge: 45.79686, -122.23703
B - Adit: 45.78765, -122.20918
C - Trailhead camp: 45.78486, -122.21427
D - Miners Queen: 45.77551, -122.19149
Best season: Late summer
Land manager: Gifford Pinchot National Forest
Material: Fine gold, small flakes, black sands
Tools: Pan, sluice, highbanker, dredge
Vehicle: Any; 4WD suggested
Special attraction: Moulton Falls
Accommodations: Sunset Falls CG; dispersed camping on Copper Creek
Finding the sites: From I-5 take exit 11 and go east on WA 502 to Battle Ground, about 6.2 miles. Turn left onto WA 503 and drive 5.6 miles, then turn right onto NE Rock Creek Road. It soon becomes NE 152nd Avenue, which you take for 1.2 miles to Lucia Falls Road. Go 7 miles to Sunset Falls Road and drive 7.4 miles to Sunset Falls Campground. Here it gets tricky—you actually have to enter the campground to access Copper Creek. Go south to a bridge, then go left onto Sunset Falls Lane, which becomes Sunset Hemlock Road (NF 41). Drive 3.1 miles and look for an impossibly sharp right turn onto NF 4109. It is just 0.3 mile to the GPAA bridge, and there is good creek access and camping before you cross. To reach the adit, return to NF 41 and drive 0.6 mile to the junction with the Copper Creek trailhead. Skip the trailhead for now and stay left; drive another 0.8 mile. Keep an eye out for the adit on your left, right above the road. To reach the camping spot at the Copper Creek trailhead, backtrack to the same junction and take the left turn headed down the hill about 0.6 mile. There is a good camping spot and turnaround at the end of the road. To reach the Miners Queen adit, you'll need to hike up Copper Creek about a mile, then bear left onto Miners Creek and go about 0.3 mile. You should see three different adits; all were active claims until recently, prone to cave-in, and full of deadly fumes, so keep out. There is another trio of adits about 0.2 mile farther up Copper Creek on the USGS topo maps, reportedly at around 45.7719, -122.2019.

Copper Creek contains plenty of bedrock and large boulders to work around for gold and black sand concentrates.

Prospecting

The GPAA claim at Copper Creek is very popular, but even if you are not a member, you should check it out. There is bedrock to inspect for traps and cracks, and moss on the rocks, and you are close to the source of the gold up on Miners Creek. You might find enough colors to convince you to join the club! Up at the adit you can grab a bucket of mineralized earth and take it back to the creek to pan, but it is sticky and can make a mess of your car's interior. The trailhead at Copper Creek is an excellent camping spot, right on the creek bank, although like most of these spots in the Cascades, it can be a bit damp in the rainy season. Bring a big tarp to keep your fire pit dry. Finally, the hike up to Miners Creek and upper Copper Creek is easy, although there is a ford involved toward the end. That's why this site works better toward the end of summer, when the water is lowest. Note that if you look on a map, you are only about 3 miles northwest of the Prospector Creek locale on the Washougal River—this is all part of the same mineralized zone, documented in the Washington Department of Natural Resources Information Circular 60, *St. Helens and Washougal Mining Districts of the Southern Cascades of Washington* (1977), by Wayne S. Moen. This area is more noted for its copper deposits, but fortunately there is enough gold to make it interesting for gold panners.

28 Lewis River

See map on page 64.
Land type: Creek, river
County: Skamania
GPS:
A - Sunset Falls CG: 45.81808, -122.25175
B - Texas Gulch: 45.82363, -122.16456
Best season: Late summer
Land manager: Gifford Pinchot National Forest
Material: Fine gold, small flakes, black sands
Tools: Pan, sluice, highbanker, dredge
Vehicle: Any; 4WD suggested for Texas Gulch
Special attraction: Moulton Falls
Accommodations: Sunset Falls CG; dispersed camping along the river above the campground
Finding the sites: From I-5 take exit 11 and go east on WA 502 to Battle Ground, about 6.2 miles. Turn left onto WA 503 and drive 5.6 miles, then turn right onto NE Rock Creek Road. It soon becomes NE 152nd Avenue, which you take for 1.2 miles to Lucia Falls Road. Go 7 miles to Sunset Falls Road and drive 7.4 miles to Sunset Falls Campground. To reach the mouth of Texas Gulch, go straight past the campground on what is now NF 42 for 5 miles to the trailhead.

Prospecting

Back during the Great Depression, the McMunn Placer, the Lewis River Placer, and the South Fork Lewis River Placer all recovered placer gold. None of these operations stayed active for long, but there are still good colors and ample black sands if you can get to a good crack on bedrock and clean it out. I have had good luck cleaning moss from the rocks here as well. Local prospectors have reported good colors from Moulton Falls Regional Park at 45.8317, -122.3889. Sunset Falls Campground at Site A is a good central base for working both Copper Creek and the Lewis River, and you can access the water and pan in the campground. However, the river runs almost straight through here, so you will do better closer to the minor falls upstream or at the bend downstream from the bridge. Ideally, you would want to explore

Check garage sales and thrift stores for common kitchen implements like strainers that could end up in your panning "go bag."

the mouth of Copper Creek, but that is on private land. The Texas Gulch locale at Site B is intriguing; you can pan colors where Green Creek merges with the East Fork of the Lewis River. However, the old placer workings at Texas Gulch are much farther up that trail, probably at 45.8049, -122.1327, at least 2 miles. I have been rained out twice and snowed out twice trying to mount that expedition, and even wondered if it would be easier to come in from the south via NF 41 and the old Fourth of July Camp near Lookout Mountain.

29 Cape Disappointment

See map on page 64.
Land type: Ocean beach
County: Pacific
GPS:
A - Camp Loop A: 46.28974, -124.07612
B - Beards Hollow: 46.30343, -124.07421
Best season: Any
Land manager: Cape Disappointment State Park
Material: Fine gold, black sands
Tools: Pan, bucket
Vehicle: Any; 4WD suggested for Site B, requires beach drive
Special attraction: Fort George Brewery in Astoria
Accommodations: Cape Disappointment State Park
Finding the sites: From Astoria drive north across the Columbia River on US 101 for 16.6 miles to Spruce Street in Ilwaco. Turn left onto WA 100 South and drive 3.8 miles, first paying the required state park fees at the front gate, then proceeding to Site A. Find the loop closest to the ocean; there is parking on the left. To reach Beards Hollow, drive north on US 101 from Ilwaco about 1.8 miles toward Seaview, then turn left (west), and head to the beach, about 0.6 mile, then 1.9 miles south to the end of the road.

Prospecting

There was a time when you could see a large group of prospectors working the sands at Site A, using battery-powered panning machines. Unfortunately, such machines use Jet-Dry or other automatic dishwashing solutions to keep surface tension under control, and the latest restrictions banned such machines. The gold is so fine that it can attach to oils and ride the water without the chemical, which biologists fear would damage wildlife. Nowadays you're more likely to see prospectors haul out a couple buckets of black sands to process at home. The best approach is to take the trail to the sandy beach, move a dozen yards off to the southern side, and dig a hole about 6 feet across and 4 feet deep. The dry sand will try to cascade down, so try digging in tiers. You'll see horizons of black sands, some with streaks as much as a foot thick.

Destroyed entrance to one of the old mines on Camp Creek

You can sample as you go down to see which streaks are better or eyeball the sands for the darkest material. It is surprisingly heavy to shovel. Note that it is *crucial* that you fill your hole back in, so keep this in mind as you move material to the side. The Beards Hollow locale at Site B also concentrates black sands, and nearby creeks that enter the ocean are good sources of black sands too. The beach drive is best attempted only with 4WD.

NORTHEAST WASHINGTON

30 Sullivan Creek

Land type: Creek
County: Pend Oreille
GPS:
A - Bridge: 48.83926, -117.26665
B - Campgrounds: 48.83303, -117.24575
C - Access: 48.83715, -117.21516
D - Upper: 48.90128, -117.08317

Northeast Washington A

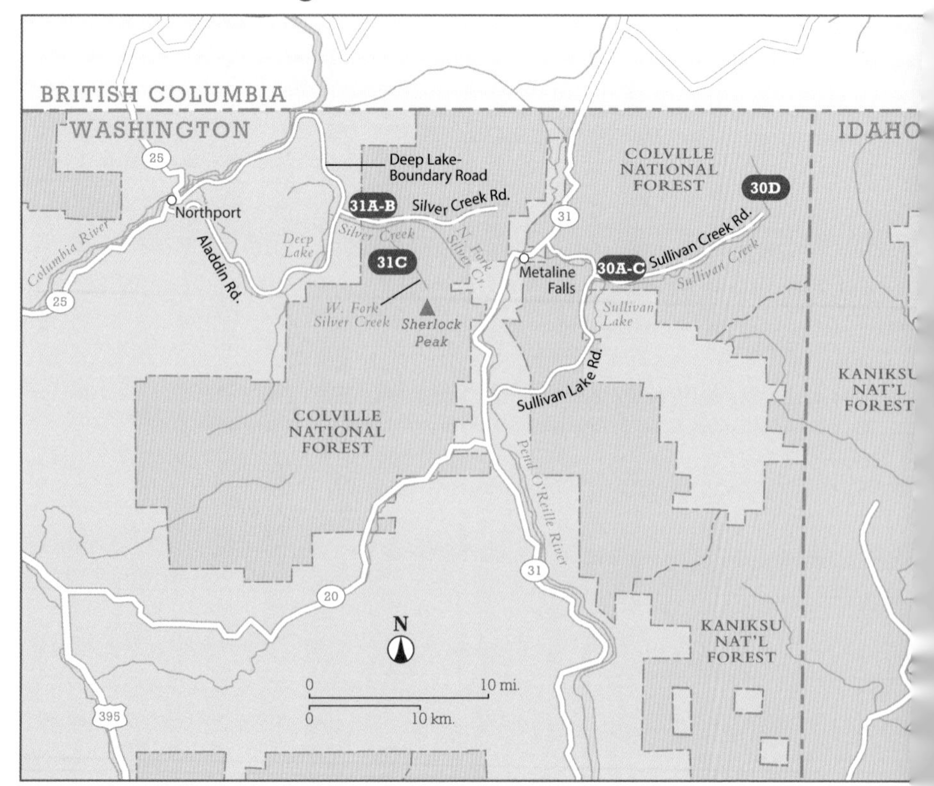

Panning at the public area on Sullivan Creek, an area noted for coarse gold and plentiful colors

Best season: Late summer
Land manager: Colville National Forest
Material: Fine gold, coarse flakes, black sands
Tools: Pan, sluice, highbanker, dredge
Vehicle: Any; 4WD suggested if you go higher up
Special attractions: Sullivan Lake millpond, Boundary Dam, Gardner Cave
Accommodations: Campgrounds on Sullivan Lake and along lower Sullivan Creek; dispersed camping all along Sullivan Creek
Finding the sites: From Metaline Falls drive north on WA 31 for 2.3 miles to Sullivan Lake Road and turn right. Follow Sullivan Lake Road 4.8 miles, then turn left onto NF 22. If you miss this left turn, you will end up back in Ione, so if you reach the lake, you have gone too far. Follow NF 22 for 1.1 miles to reach the bridge. Go another 1.1 miles to reach the camping area along the creek. There is a great access point another 1.6 miles up, and the final locale is 8.2 miles farther, although you'll pass plenty of good-looking places to pull over and test.

Prospecting

This is an open area set aside for panning and prospecting—one of the few left in the Pacific Northwest. We found consistently good color except for one spot that turned out to be dredge tailings. The gold gets noticeably coarser as you go upstream, and there are numerous places to test. I have not visited any of the lode mines up here, but some of them were copper prospects, with gold just a by-product. There are even tungsten locales up here—the mineralization is crazy. Two words of warning about camping up here: (1) There are some ferocious swarms of biting flies, and they drove us crazy one morning. We ended up breaking camp in record time and fleeing to a cafe in Metaline Falls. (2) There are some big-eared camp pack rats that prowl the fire pits when all the campers have retired. We woke up late one night to the sound of three furry critters rustling through the garbage and making a terrible racket. This is also bear country, to make matters even more interesting. Still, there is mineralization throughout these hills, and along the way to Boundary Dam and Gardner Cave, farther north, there are a couple of easy mines to check out: the Oriole Mine, at 48.8601, -117.41386, and the Josephine Mine, at 48.8789, -117.3783, for starters.

31 Northport

See map on page 72.
Land type: Creek
County: Stevens
GPS:
A - Lower bridge: 48.90469, -117.57759
B - Upper bridge: 48.90969, -117.53819
C - Mines: 48.88471, -117.53361
Best season: Late summer
Land manager: Colville National Forest
Material: Fine gold, black sands; silver and lead
Tools: Pan, sluice, shovel, bucket
Vehicle: Any; 4WD suggested
Special attraction: Northport
Accommodations: Services in Northport and Kettle Falls. Beaver Lodge CG near Gillette Lake; dispersed camping on USFS land and on Gladstone Mountain and Red Top Mountain.
Finding the sites: From Northport drive northeast 9.9 miles on WA 25/Northport-Boundary Road toward Canada, then swing off just before the border onto Deep Lake Boundary Road. Head south to just past Leadpoint, about 8.7 miles, and locate Silver Creek Road. Drive 0.6 mile to the junction with Gladstone Road to reach the first spot. (***Note:*** You can take Gladstone Road up the mountain and possibly access the mines at the top of the mountain. We had to turn around.) Drive 1.7 miles farther up Silver Creek Road, then turn right onto NF 075 and go 0.3 mile to the second bridge. Continue on NF 075 up Gladstone Mountain for 4.3 miles to reach a large selection of mines, adits, shafts, and prospects.

Prospecting

If not for Grand Coulee Dam and the creation of Franklin D. Roosevelt Lake, the Northport area would be a hot zone for placer mining. Before the lake inundated this stretch from Kettle Falls to the Canadian border, there were dozens of operations at work on the Columbia River, such as at Goodeve Creek Bar, Northport Bar, Bossburg Bar, and Valbush Bar, just to name a few. This area is remote, sparsely inhabited, and loaded with mineralization,

One of the many cliffs worked at Gladstone Mountain. This Lexus SUV had 10-ply tires, which helped a lot, but lacked an adequate suspension and bottomed out a couple times on the drive up. We replaced it soon after with a Jeep.

although many of the mines explored silver, lead, iron, and zinc deposits. Still, gold was a valuable by-product, and there are hundreds of named mines and prospects in these hills. This Gladstone Mountain area is just a taste of the heritage you will find up here if you have time to explore. Silver Creek offers plentiful black sands and occasional chunks of silvery metal, so it is very rewarding. Be aware of patented land at the top of Gladstone Mountain, and don't be surprised by gates on roads that you really wish were open for exploration. Fortunately there are very few active claims here. Site A is a good camping spot and is right on the water. Site B offers access to the water as well.

32 Molson

Land type: Museum
County: Okanogan
GPS: 48.97592, -119.19985
Best season: Late summer
Land manager: Private
Material: Mining artifacts
Tools: Camera
Vehicle: Any
Special attractions: Hee Hee Stone, Pflug Mansion
Accommodations: Beth Lake CG

Northeast Washington B

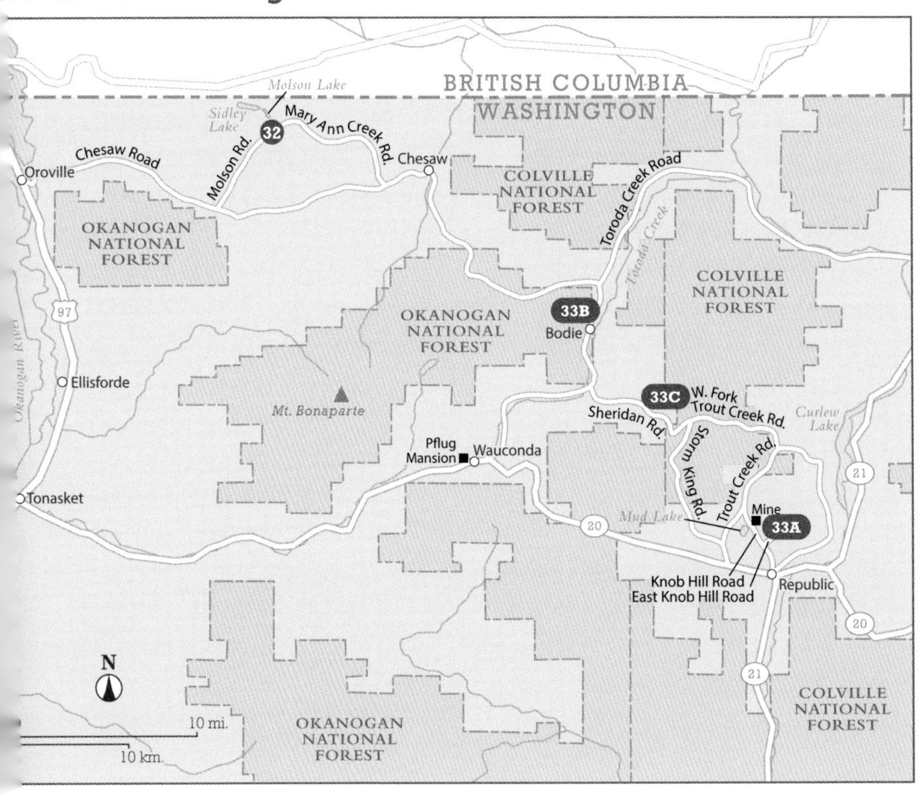

The open-air museum at Molson is family friendly, educational, and well worth the drive.

Finding the site: Old Molson is the site of a fun outdoor museum that parents can use to let their kids blow off some steam. From US 97 in Oroville, drive east on Central Avenue several blocks to Cherry Street, which becomes Chesaw Road. After 9.1 miles turn left onto Molson Road and go 5 miles to the museum parking area. From Bodie head north on Toroda Creek Road for 2.1 miles, turn left onto Chesaw Road, drive about 15 miles to Chesaw, then another 1.5 miles to Fields Road. Turn right and drive about 4.7 miles to Molson.

Prospecting

We had high hopes for Mary Ann Creek when we scouted this area, but there just does not seem to be any access to the creek bed, and there wasn't much water either. Most of the mineralization at Buckhorn Mountain is unavailable as well. We stopped here and let the kids run around and tire themselves out while we snapped some pictures at the old assay office and took a break.

33 Republic

See map on page 77.
Land type: Creek
County: Ferry
GPS:
A - Knob Hill: 48.67086, -118.75036
B - Bodie: 48.83455, -118.89692
C - Zalla M: 48.75956, -118.83259
Best season: Late summer
Land manager: Colville National Forest and Okanogan National Forest
Material: Fine gold, small flakes, black sands
Tools: Pan, bucket, heavy hammer
Vehicle: Any; 4WD suggested for Zalla M
Special attraction: Stonerose
Accommodations: Sanpoil River and Beth Lake CGs; dispersed camping at Zalla M
Finding the sites: Republic sits on WA 20, more or less halfway between Tonasket and Kettle Falls. From the top of Clark Avenue in Republic, turn left onto Knob Hill Road and drive 1.4 miles, then turn right onto Knob Hill East and drive about 0.4 mile to a good vantage point to see the old mine site. The Zalla M Mine on Horseshoe Mountain requires a sturdy 4WD. Drive back down Knob Hill East Road to the main Knob Hill Road and go past the mine to Trout Creek Road, about 1.9 miles. Take Trout Creek Road for 3.7 miles, then turn left onto West Fork Trout Creek Road. Stay on the main road for 7.9 miles to a nice camping area; there is a big open-pit mine with lots of quartz crystals another 0.1 mile up this road. To reach Bodie, any vehicle will do, so if you skip the Zalla M, start from Republic. Head east on WA 20 for 16.4 miles to Toroda Creek Road, then drive 11.4 miles north to the ruins along the road. (From here you could continue to Molson.) You can reach Bodie from the Zalla M by continuing north on West Fork Trout Creek Road for 0.3 mile and then heading west, down the hill, on Sheridan Road. Follow it for 4.3 miles; it will briefly become Cougar Creek Road, then hook up with Toroda Creek Road. Bodie is just 4.5 miles north.

Prospecting

The southwest slope of Klondike Mountain contains multiple prospects, mines, adits, and diggings if you want to explore out here. The same goes for

This old log cabin is right along the road at Bodie, not far from Republic.

the southwest corner of Republic, above Granite Creek along Pendy Road. There may be access to Granite Creek at 48.6427, -118.7419, via Airport Road and Lilly Creek Road, but I was not able to verify it. Most of the creeks in this general area contain gold, so if you encounter a dry gulch, consider using a small shovel and a hand broom to collect a sample you can pan later. You might stumble on very coarse gold or gold still bound up in quartz. I panned colors on the Sanpoil River, far south of Republic. The Knob Hill Mine is not open for tours, and the tailings are not very noteworthy, but it's worth checking out just to get a sense for the scale of the gold extracted from the Republic area. There used to be a prospecting shop in Republic where you could get supplies and updated information, but it closed. The hardware store sells pans and shovels, and the counter help there does have good advice. The Stonerose Interpretive Center at Sixth and Kean sells several books about local mining, and if you have any fossil diggers in the crew, it is worth a stop. At the Zalla M site, there are multiple prospects worth checking out, although the one mine shaft there is a bit concerning for parents. If you take the back way to Bodie from the Zalla M, look for access to Cougar Creek as you near the bottom, but it may all be private. We panned colors from Toroda Creek behind Bodie.

HONORABLE MENTIONS

The following Washington locales did not make it into the book, but they're worth a visit if you're in the area.

A. Cayada Creek

This spot is north of Mount Rainier National Park, and it was never a big producer. I sampled Cayada Creek in three places and saw one color, so I skipped it. The mine itself is pretty cool but not enough to bump any of the other spots. I tried the Carbon River and did not have much luck with it either.

B. Cle Elum

This one just got away from me, and I still need to check it out. I have spent a lot of time around Liberty, but I have not prospected much above Roslyn, so it's on my to-do list.

C. Columbia River

Unless the Columbia drops way, way low, you just aren't going to have much luck out there. I found a few places on BLM land around Roosevelt, but you have to dodge railroad tracks and private land.

D. Conconully

The old mining district at Conconully was a major producer back in the day, but it is mostly private now. I panned a few colors from the North Fork Salmon Creek, but I need a knowledgeable local guide to show me around up there. I had high hopes for the Mineral Hill area, but I ran into a gate. I had much better luck at Ruby and the Last Chance Mine, farther south.

E. Foss River

Like Miller River, the Foss River district has never been very kind to me; Money Creek is much better. The water seems to come rushing off the hill-sides with far too much energy to trap any black sands or gold particles, and it does not seem like there was really that much to begin with.

F. Granite Falls

I finally found the old placer area at Granite Falls by hiking on the other side of the road from the walk down to the falls provided by the power company. I found an old, rotten ladder that spanned a landslide and probably once reached a nice beach area with great bedrock traps. Only someone with a death wish or a crazed kayaker who feels invincible could get in there now. It is just not possible without jet packs, hovercraft, or a transporter.

G. Klickitat River

Far, far up the Klickitat, there is a very small placer on Surveyors Creek. Or at least there was at one time. The Yakima Indian Reservation requires a permit to penetrate the reservation that far up, and the joke is that nobody has ever been granted a permit in the thirty-five years that the geologist up there could recall. There is a pair of nice campgrounds south of the reservation at Liedl Park and Stinson Flats, and they are loaded with black sands. I recovered a few colors there, but that was it.

H. Miller River

Unlike nearby Money Creek, the gravels at Miller River have never been very kind to me. I have panned a few colors out of the mossy boulders, but I always end up feeling guilty about scalping the rocks and have given up on Miller River.

I. Mount Rainier

During the late summer months, there are several areas around the northeast corner of Mount Rainer worth exploring. The Crystal Mountain ski resort is drained by Silver Creek, and I panned some decent colors from it at the horse camp at 46.9609, -121.4815. There were several more prospects along the ski slopes. On the other side of the ridge to the southeast, along the American River just off WA 410, there's a pair of mines starting at 46.9024, -121.4326. I found an old camp up Morse Creek at 46.9087, -121.4428, but the colors were not that great.

Part II: Oregon

Oregon

WASHINGTON
PACIFIC OCEAN
OREGON
IDAHO
NEVADA
CALIFORNIA

SIUSLAW NAT'L FOREST
MOUNT HOOD NAT'L FOREST
WILLAMETTE NAT'L FOREST
DESCHUTES NAT'L FOREST
OCHOCO NAT'L FOREST
UMATILLA NAT'L FOREST
WHITMAN NAT'L FOREST
WALLOWA NAT'L FOREST
MALHEUR NAT'L FOREST
UMPQUA NF
WINEMA NATIONAL FOREST
FREMONT NAT'L FOREST
SISKIYOU NAT'L FOREST
ROGUE RIVER NAT'L FOREST

Vancouver
Portland
Hillsboro
Beaverton
Gresham
Tualatin
Oregon City
Newberg
Wilsonville
McMinnville
Woodburn
Dallas
Salem
Albany
Corvallis
Lebanon
Eugene
Springfield
Coos Bay
Roseburg
Sixes
Grants Pass
Central Point
Medford
Ashland
Klamath Falls
City of the Dalles
Redmond
Bend
Hermiston
Pendleton
La Grande
Greenhorn
John Day
Ontario
Caldwell

N
0 50 mi.
0 50 km.

BLUE MOUNTAINS (OREGON)

Blue Mountains

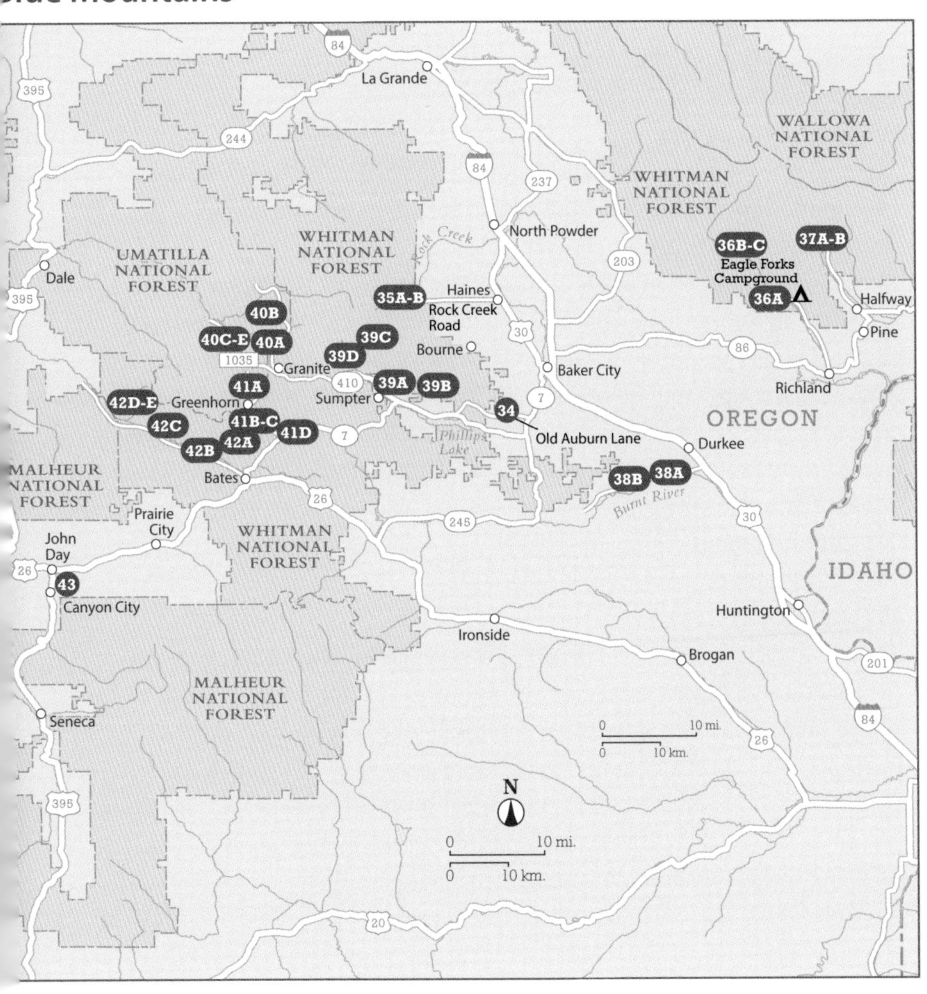

34 Auburn

Land type: Creek
County: Baker
GPS: 44.69951, -117.94811
Best season: Spring–early summer
Land manager: Wallowa-Whitman National Forest, BLM–Baker
Material: Fine gold, small flakes, black sands
Tools: Pan, sluice
Vehicle: 4WD required
Special attraction: Mason Dam
Accommodations: Developed camping along Mason Dam; primitive camping farther out Auburn Lane on USFS land
Finding the sites: From Baker City head southwest onto OR 7 about 7 miles and look for Old Auburn Lane on the right. Follow it through the ranches and be on the lookout for bald eagles resting on telephone poles. Continue 4.4 miles, then make a slight right turn going downhill, on a very rough road. After about 0.3 mile, you will reach the creek crossing; you can go farther or stop here. The Auburn cemetery is about 0.2 mile farther. This was a major placer operation back in the 1860s. The old townsite itself is about 500 feet southeast of this spot, but there is very little left there.

Prospecting

Auburn was the site of the first big gold rush in eastern Oregon, and at one time the town hoped to become the county seat for Baker County. However, there was never enough water out here, and the surface diggings petered out fast. You can find enough water to pan here in the spring and very early summer, but there are claims to dodge as of 2022. The later it gets in the summer, the harder it is to find enough water to do much, at any rate. We panned some colors at the bottom of Blue Canyon on the Powder River around 44.6783, -117.8663, and the Forest Service used to advertise the Powder River Recreation Area as open for panning at 44.6698, -117.9756. If you like this area, there is more exploring you can do. Nearby Elk Creek hosted some good placer mining at 44.7244, -117.9292. You can reach it via about 1.7 miles of rough road from Auburn, or come in from OR 7 via Elk

A pair of bald eagles guards the road up to Auburn.

Creek Lane. Farther north the Washington Gulch Placers operated at 44.7624, -117.9168, and north from there, the Nelson Placers operated out Salmon Creek Road near 44.7957, -118.0019. There is a lot of private land to avoid around these old locales, and some active claims, but the farther up you get, the closer you are to public land, so keep heading up, as this area can be fun to explore. During the wetter months, these roads become very treacherous, so you might want to stick to summer and bring home buckets of material to wash there.

35 Rock Creek

See map on page 85.
Land type: Creek
County: Baker
GPS:
A - Access 1: 44.89639, -118.08548
B - Access 2: 44.89208, -118.09435
Best season: Late summer
Land manager: Wallowa-Whitman National Forest, BLM–Baker
Material: Fine gold, small flakes, black sands
Tools: Pan, sluice; good luck getting equipment down to the water
Vehicle: 4WD required
Special attraction: Haines Eastern Oregon Museum
Accommodations: Dispersed camping on USFS land as soon as you reach the first site
Finding the sites: From I-84 take exit 285 at Powder River and head for Haines on US 30. After about 8.3 miles, take Fourth Street in downtown Haines and drive west about 0.5 mile. Continue onto Rock Creek Road for 1.2 miles, keep left to jog onto Pocahontas Road, then take the first right onto South Rock Creek Lane. Follow it for 2.2 miles, then swing right to stay on it and avoid Willow Creek Lane. Go 4.4 miles as this road becomes NF 5520 to reach the first access spot, which has a nice camping area. The next spot is about 0.6 mile farther. If you can make it 2.8 miles from the first site, you will reach Eilertson Meadow, with a fork in the road to old, private mines such as the Chloride and Highland-Maxwell.

Prospecting

The disintegrating granite of the high Elkhorn Mountains supplies lots of black sands, mostly magnetite, so expect to battle that. You will also see a lot of brassy muscovite mica up here. However, there is good gold in Rock Creek—so much that farther down, at Haines, the roads would glitter after spring floods delivered fresh loads each year. You won't have much luck trying to reach bedrock, but if you can move some big rocks around, you should do well. The mines aren't far away, and you could spot quartz with enough rusty staining to merit breaking apart. Be aware of old family cabins and

Rock Creek is far above Haines, Oregon, and offers access to some very old lode mines and prospects.

mines up here; we don't mind if you take a sample from the tailings piles at the Highland-Maxwell, for example, but leaving trash, vandalizing structures, and committing crimes just wrecks things for everyone else. Pack out your garbage, use a shovel to bury waste, and be a good citizen even when nobody is watching.

36 Eagle Creek

See map on page 85.
Land type: Creek
County: Baker
GPS:
A - Eagle Forks CG: 44.89035, -117.26098
B - Eagle Flat: 44.97098, -117.35934
C - Camp: 45.00155, -117.41154
Best season: Late summer
Land manager: Wallowa-Whitman National Forest
Material: Fine gold, small flakes, black sands
Tools: Pan, sluice, highbanker, dredge
Vehicle: Any; 4WD suggested—roads are rough in places.
Special attraction: *Paint Your Wagon* movie set locale at 45.0243, -117.3361
Accommodations: Developed camping at Eagle Forks and Two Color CGs; primitive camping all throughout area, with good access to the water easy to find
Finding the sites: From just west of Richland on Main Street/OR 86, look for New Bridge Road and head north. After about 2.3 miles, you will find New Bridge and a left on Sparta Lane that will take you to that old mining area. There's no water up there to pan with, and just crumbling structures. Stay on this road as it becomes Eagle Creek Road and go another 5.1 miles. Now this main road becomes NF 7735. After 2.4 miles you will see signs and a big cattle guard as the side road leads down to the campground at Eagle Forks. To reach Site B at Eagle Flat, stay on the main road leaving Eagle Forks Campground for 3.6 miles until it bears right and becomes NF 7720. After 1.3 miles stay left to take NF 77. Follow this road for 8.7 miles; the last 2.5 miles are along Eagle Creek. To reach the mouth of East Eagle Creek, continue about a mile. (NF 7745 follows East Eagle Creek and will take you up to the *Paint Your Wagon* site.) The final site is 2.9 miles farther up Eagle Creek Road; Two-Color Campground is even farther up. Note that you can come in via Medical Springs on Big Creek Road or via Sparta if you have good maps, a sturdy vehicle, and GPS.

Prospecting

This area is picturesque, with the Eagle Caps towering above you and bright white granite boulders in the rushing creeks. There is lots of black sand and

While in the area, head up East Eagle Creek and check out the locale where Hollywood producers filmed the movie *Paint Your Wagon* starring Clint Eastwood and Lee Marvin in 1968.

mica to contend with, but there is also enough gold to make things interesting, and there's no shortage of water. The area is noted for marine fossils in limestone, too, with a significant ichthyosaur fossil discovered here near the mouth of East Eagle Creek. Years ago, Eagle Forks Campground at Site A was set aside as a public panning area, but too many miners cut into the banks and the Baker office stopped advertising it as open. The first time we camped there, we walked up Eagle Creek via the trail and panned good colors there. After that we just panned on the bend, moving boulders. We dragged a magnet through the dusty soil around the campsite and it was soon thickly coated with black sands. The Eagle Flat area at Site B hosted a large placer mine at one time, but there is not much bedrock showing here. This is another good place for a central camp to mount daily recon tours, and was unclaimed as of 2022.

The mouth of East Eagle Creek has very good color, as there were mines above, both on East Eagle Creek and along Eagle Creek itself. The Hillsboro Gold Mining Company had a big operation here. There are claims to dodge throughout this drainage and a network of roads leading down to the final camp locale at Site C, with concrete flooring still in place and quite a few rusty nails in the soil. This was the Yellow Nugget Placer Mine. You could spend a week up here exploring old sites like Sparta, Sanger, and Cornucopia. There is no water at Sparta, as I mentioned, and Sanger is down to one leaning cabin. If you do intend to explore, bring a big, heavy hammer to break up any large chunks of rusty-stained quartz you find. You might get lucky and find native gold inside.

37 Cornucopia

See map on page 85.
Land type: Mountainside, creek
County: Baker
GPS:
A - USFS corner: 44.94529, -117.18132
B - Union Mine: 45.01241, -117.21451
Best season: Late summer
Land manager: Wallowa-Whitman National Forest, BLM–Baker
Material: Fine gold, small flakes, black sands; pyrites and sulfides in quartz
Tools: Pan, heavy hammer
Vehicle: 4WD required for Union Mine; decent road to Cornucopia, then gets very rough
Special attraction: Hells Canyon
Accommodations: McBride CG; dispersed camping on USFS lands
Finding the sites: From Baker City drive east on OR 86 about 52 miles to reach the Halfway Spur, which takes you into Halfway after about a mile. Continue onto Halfway-Cornucopia Highway and drive about 6.3 miles to the start of USFS lands; this is Site A. To reach the Union-Companion vein of the Cornucopia Mine, continue into what is left of Cornucopia (mostly cabins at this point) and stay left. The road turns back around to the left and heads south, past more old buildings. After about 0.3 mile from the tight left turn, a rough road heads right, uphill. Less than 0.25 mile from the right turn, you should be able to locate Jim Fisk Creek, as it runs right along, and apparently sometimes in, the road at this point. But it is hard to tell, as it is usually dry. After another 1.1 miles the road leads to the disintegrating ruins of the major mine, with a large tailings pile to explore.

Prospecting

Cornucopia, which means "horn of plenty," was a major producer right up to World War II, when gold mining became unimportant to the war effort. The mine never reopened after the war, although it has major reserves left to exploit. The Union-Companion and the Last Chance were the major mines up here, and the Union-Companion vein crops out on the surface for more than a mile to the northeast. The Coulter Tunnel is not far from the big left

Foundations for one of the structures near the Union Mine, on Jim Fisk Creek above Cornucopia

turn toward Jim Fisk Creek and sees activity during the summer, so use discretion in that area. Pine Creek was a major placer producer, but bedrock lay under as much as 60 feet of overburden. Still, it is easy to spot major tailings along Pine Creek as you drive up. Be cautious about panning in Pine Creek—look for active claims with current signs and markers. In 2022 we field-checked and there were claims all along Pine Creek. Jim Fisk Creek is often dry by July, but you could scoop up a bucket of pay dirt and clean it where there is water. At this point you want to get your rockhounding eye dialed into the white quartz and look for gray or rusty streaks and staining. The gold here is associated with pyrite and sulfides, with greenish malachite, a copper sulfide, especially easy to spot. You can further explore the hills, but they are very steep and hot in the summer. On the other side of the draw from the Union vein, West Pine Creek leads up into the mountains and access to more mines, but the best way to try it is via packhorse from the Cornucopia Lodge (cornucopialodge.com). The lodge offers trips to various panning locales around Cornucopia. Note that this is one of the more reliable areas for spotting small black bear cubs, which invariably have a mama bear nearby.

38 Burnt River

See map on page 85.
Land type: River
County: Baker
GPS:
A - Lowest: 44.57041, -117.59042
B - LDMA camp: 44.56125, -117.61346
Best season: Late summer
Land manager: BLM–Baker, Wallowa-Whitman National Forest
Material: Fine gold, small flakes, black sands
Tools: Pan, sluice, highbanker, dredge
Vehicle: Any; 4WD suggested—gravel roads are rough and contain troublesome washboards in places.
Special attraction: Unity Reservoir
Accommodations: Dispersed camping only along this stretch of the Burnt River
Finding the sites: From Baker City drive south on I-84 about 22.5 miles to Durkee, at exit 327. Turn west and drive 0.4 mile on Durkee Road, then head north 1.6 miles on old US 30 to Burnt River Canyon Lane. Site A is about 7.1 miles up the river. The LDMA camp is 1.6 miles farther up the river.

Prospecting

This part of Baker County has seen a lot of gold-mining activity, and it continues to the present day. The Hoffman family worked here one summer for the *Gold Rush* show on the Discovery Channel. The Burnt River stretches beyond here, past Durkee, to the mouth of the river on the Snake, just below Huntington. I have rock-hounded to the mouth of the river, and there are good black sands and decent colors there. The stretch along I-84 is mostly private, but I worked a claim with longtime prospector Kevan Reedy near the concrete plant, and there are interesting mines on Shirttail Creek, up Sisley Creek, and all the way up Jordan Creek to the Bassar Diggins Campground. Plus, the Lost Dutchman's Mining Association (LDMA) has another site at the Weatherby exit. That entire area is worth exploring, but be aware of private property and active claims.

New wash plant on Clarks Creek, just above the Burnt River canyon near Bridgeport

This lower Burnt River stretch is usually open, but watch for claim markers, as things do change. Consider checking in at the LDMA camp on the Burnt River at Deer Creek for up-to-date info, and maybe watch them working their wash plant if they are active. Like most places, the gold out here resides in bedrock cracks, under boulders, on inside bends of the river, and at the head of forming gravel bars, and you will need a deep hole. One issue with the river is that with so many cattle ranches at Hereford and Bridgeport, the water can run very brown by August. If that is the case, try exploring up Clarks Creek, all the way to Mormon Basin and Rye Valley, and check for open areas around there. Some of the major creeks running south into the Burnt River, such as Auburn Creek, McClellan Creek, and especially Pine Creek, are also interesting. You need a rugged vehicle for that kind of work, and it gets hot out there, but it is a nice way to spend a day. If nothing else is working, try heading up the Burnt River via CR 529 to Greenhorn; there's a good spot at 44.58367, -118.2299.

39 Sumpter

See map on page 85.
Land type: Creek
County: Baker
GPS:
A - Sumpter Dredge: 44.74189, -118.20328
B - Deer Creek: 44.74622, -118.10429
C - E&E Mill: 44.82854, -118.19857
D - McCully Forks CG: 44.76932, -118.24796
Best season: Spring–fall
Land manager: Wallowa-Whitman National Forest
Material: Fine gold, small flakes, black sands
Tools: Pan, sluice; heavy hammer for Bourne area; metal detector for tailings
Vehicle: Any to reach dredge; 4WD suggested for exploring
Special attraction: Sumpter Railroad
Accommodations: Union Creek CG at Philips Reservoir; McCully Forks has a fee campground at the highway, but Deer Creek is now free, as it lacks tables, fire rings, or outhouses. Motels and B&Bs in Sumpter and Baker.
Finding the sites: To reach the dredge, drive south and then west from Baker on OR 7 for 25 miles, then turn right onto OR 410 and drive 2.9 miles. As you reach Sumpter, you will spot an open-air museum that is fun to tour and then locate signs for the Sumpter Valley Dredge State Heritage Area. You will see the dredge before you turn toward it on Austin Street. To reach Deer Creek, turn north from OR 7 near McEwen, about 22 miles west of Baker. Deer Creek Road takes you to the spot, as does Larch Creek Road, which comes in more from the east. Take Deer Creek Road (NF 6550) about 3.3 miles, then turn right onto NF 6530 and drive another mile. You will see the old camp. To reach the old E&E Mill at Bourne, drive north from Sumpter about 0.6 mile toward Granite, then turn right onto Cracker Creek Road. Drive 6.2 miles, past the cabins, then turn left onto NF 5505 and go about 0.25 mile. Finally, to reach McCully Forks, drive from Sumpter on OR 410/ Granite Hill Road about 2.8 miles, then turn right onto McCully Fork Road and drive past the campground. Go about 0.3 mile or keep going to the end of the road.

This picture of the Sumpter Dredge dates to 2007, before volunteers repainted it and before trees grew up around it.

Prospecting

If you have never panned before or just want a refresher, the folks at the Sumpter Dredge offer panning lessons throughout the day during the summer, and they will let you pan anywhere in the state park, so you can work in the tailings in the shade. The gift shop is excellent (although they *still* don't stock this book!) and the staff is incredibly helpful. There are tours of the dredge as well, and railroad excursions into the dredge's tailings field near McEwen. I've heard legends about giant gold nuggets supposedly still out there that the dredge could not capture, and metal detectorists regularly haunt the tailings piles and ponds throughout this area. Both Site B at Deer Creek and Site D at McCully Forks Creek used to be advertised by the US Forest Service as areas open to panning, and they were hammered pretty well until the program was discontinued. You will still see literature that also lists sites at Antlers (upper Burnt River), Eagle Forks (on Eagle Creek), and Powder River (below Mason Dam). I have panned colors at the E&E Mill locale at Site C, and when I first visited the mill in the 1970s, there were old drill samples all over the place where you could see the pyrite mixed with rusty-stained quartz. Cracker Creek is mostly claimed now and has seen a bit of a revival; the remains of the dredge that dug up the area are in a pond at 44.7652, -118.2026. Most of the major roads out here wind through mining country, and you may have as much fun with a hammer as with a pan.

40 Granite

See map on page 85.
Land type: Creek
County: Grant
GPS:
A - New York Mine: 44.84789, -118.40249
B - North Fork CG: 44.91267, -118.40126
C - Buck Creek: 44.8414, -118.4945
D - Access: 44.8393, -118.5068
E - Trailhead: 44.8462, -118.5178
Best season: Late summer for low water
Land manager: Wallowa-Whitman National Forest
Material: Fine gold, small flakes, black sands
Tools: Pan and shovel, mostly; larger equipment at Tabor Diggings and below is OK.
Vehicle: 4WD suggested for exploring
Special attraction: Granite ghost town
Accommodations: Developed campsite at North Fork of the John Day; primitive camping along Granite Creek
Finding the sites: Granite is about 16 miles northwest of Sumpter; use the Granite Lodge as a staging area, but services there are intermittent, and you should probably gas up in Sumpter. From the main Granite intersection of Granite CR 24 and NF 73, turn north on NF 73 and proceed about 3.1 miles. The ruins of the New York Mine are on the left. Drive another 5.5 miles north to reach Site B, the campground at the North Fork of the John Day River. To reach more Granite Creek sites, return to Granite and drive west on CR 24 about 1.4 miles, then turn right (north) onto NF 1035. You will see the tailings immediately. Drive 1.3 miles past some private ranchland to reach the Tabor Diggings, a rich placer mine in the past. It is located where Ten Cent Creek drains into Granite Creek, but it is under claim. To reach the rest of the sites on Granite Creek, continue west about 3 miles and look for access roads that lead south, to the water. The trailhead at Site E is about 0.2 mile off the 060 spur, and it is very rough in places.

It took a full crew to seat the barrel of the trommel on this Granite Creek placer mine.

Prospecting

The Granite Creek area boasts lode mines and placer diggings throughout the drainage. Ten Cent Creek got its name because you could easily pan ten cents' worth of gold per pan, back when gold was $16 per troy ounce. There are extensive tailings, and the easy gold was recovered long ago. Chinese miners moved in and got a lot of the difficult gold, too. There is a roadside history marker celebrating the extensive Ah Hee Diggings at 44.8286, -118.4146. Still, there are plenty of colors remaining here. Keep an eye out for private land, stream reclamation projects, and current claims. Most of the roads out here lead to current or abandoned mines, so you could have some fun afternoons just exploring with your 4WD or ATV.

The New York Mine at Site A is not open for exploration, but it's a somewhat photogenic abandoned mine; three more nearby are the Cougar Mine (44.8485, -118.4129), Buffalo Mine (44.8663, -118.3906), and Boston Tunnel Mine (44.86331, -118.39214). The town of Granite itself is also very picturesque. The North Fork Campground at Site B on the North Fork of the John Day River is quiet and scenic and offers access to the North Fork, especially gravel bars between Trail Creek and Trout Creek. While in the wilderness area, you can only pan a few sample pans, so don't expect to hunker down in one spot and make a big hole—keep moving and leave no trace. Out on Granite Creek you should do well after noting the Tabor Diggings and heading up either Ten Cent Creek or Granite Creek. There once was a big hydraulic mine at East Ten Cent Creek. Try any access spots you can find from Tabor Diggings to the trailhead. There is good bedrock lower down near the trailhead, and you do not enter the wilderness area until you head down the gated road. If you want an epic trip, hike from the trailhead to the mouth of Granite Creek where it dumps into the North Fork of the John Day, about 3.5 miles one-way. Pan samples all the way down, dodging active claims, to an enticing gravel bar at the mouth of Granite Creek, but do not plan to bring anything more than a pan, and fill in your holes completely.

41 Greenhorn

See map on page 85.
Land type: Creek, mountains, ghost town
County: Grant, Baker
GPS:
A - Ruby Creek: 44.77057, -118.49341
B - Greenhorn: 44.70926, -118.49654
C - Robinsonville: 44.71337, -118.48496
D - Porterville: 44.66831, -118.33229
Best season: Early summer for water (there's little snow)
Land manager: Wallowa-Whitman National Forest
Material: Fine gold, small flakes, black sands
Tools: Pan, sluice; heavy hammer for ore samples
Vehicle: Any; 4WD suggested—main roads are decent gravel for the most part, but with ruts and washboards. Any exploring will test your suspension and tires.
Special attraction: Greenhorn
Accommodations: Dispersed, primitive camping at Ruby Creek and down the North Fork of the Burnt River
Finding the sites: From Granite drive west on CR 24 for 3.4 miles, then swing left onto NF 13. There are many interesting lode mines near this turn, and the main road continues up to the Fremont Powerhouse and Olive Lake. You want to proceed to Greenhorn, so go 2.3 miles on NF 13 and continue onto NF 1305. To reach Site A at Ruby Creek, drive 0.6 mile on NF 1305, then turn right onto NF 1310. Drive 1.6 miles and look for a road that heads directly toward Lightning Creek and ford the creek during low water. If you are driving a good, sturdy 4WD rig and have USFS maps, you can ford the creek and even reach Greenhorn via NF 1310. Most of you will backtrack to the intersection of NF 1305 and NF 1310. To resume to Greenhorn and Robinsonville, go 4.7 miles on NF 1305, staying right. Take another right turn, onto NF 1042/Greenhorn Road at San Lou Flat, and look for a right turn at 1.6 miles. This road gets rough in a hurry, but after just 0.4 mile, you reach the old site of Robinsonville, which is Site C. Some of you might just hike in. If you skip that side trip, Site B at Greenhorn is 0.6 mile farther on Greenhorn Road. To reach Porterville from Greenhorn, drive 10.2 miles south on Greenhorn Road to the intersection with OR 7. Jog slightly left, and at 2.4 miles

Historic buildings dot the old city streets of Greenhorn.

look for parking. You can easily reach this spot in a sedan or minivan from Baker City by driving about 40 miles west on OR 7.

Prospecting

Bring a sturdy hammer with you to Greenhorn and be prepared to smash apart some rocks almost everywhere you find bedrock exposures. Sometimes you can even find chunks of quartz in the road or in veins in outcrops. There are hundreds of mines and prospects in this general area, and you will find greenstone, schist, serpentinite, argillite, granite, quartzite, and gneiss. There are garnets in many of the creeks, especially at Site A at Ruby Creek, and plentiful black sands. From Granite, once you leave Granite Creek and begin up the Clear Creek drainage, you will be frustrated by mostly private and claimed land along the creek; a crew worked up here one summer for Discovery Channel's *Gold Rush* show. Lightning Creek usually has more open access to the water. At Greenhorn check out the old town, take some pictures, grab a sample of green serpentinite, but move on, as it is private land. At Site C at Robinsonville, you will be near numerous old lode mines, such as the Owl, Red Bird, Rabbit, Ophir, Eureka, Aurora, and Humboldt, among others, but there's no water. Consider taking a sample from a dry gulch to work on at home. Finally, the old diggings at Porterville and the Whitney Placers are all claimed up as of 2022, but Site D remains open.

42 Elk Creek

See map on page 85.
Land type: Creek, river
County: Grant
GPS:
A - Vincent Creek access: 44.64289, -118.53921
B - Deerhorn CG: 44.62273, -118.58121
C - Granite Boulder Creek: 44.67984, -118.61395
D - Elk Creek: 44.71143, -118.80103
E - Armstrong Creek: 44.74318, -118.84991
Best season: Late summer for low water
Land manager: Wallowa-Whitman National Forest
Material: Fine gold, small flakes, black sands
Tools: Pan, sluice, highbanker, dredge
Vehicle: 4WD suggested unless sticking to paved roads
Special attraction: Badger Mine
Accommodations: Dispersed, primitive camping throughout this area. Deerhorn CG is your best bet for developed camping.
Finding the sites: From the intersection of OR 7 and US 26, drive north on OR 7 for 1.1 miles, then turn left onto CR 20. Drive 2.5 miles west and look for a right turn onto NF 2010. After about 2.7 miles you will reach a dirt road headed down the hill to the Vincent Creek Placers. You can continue on NF 2010 at least another 5.2 miles to check out the Morning Mine at 44.695, -118.5564. To reach Site B at Deerhorn Campground, which has some good access to the Middle Fork of the John Day, continue west on CR 20 for 3.6 miles and look for access to the left. To reach the old China Diggings at Site C on Granite Boulder Creek, continue from Deerhorn another 3.4 miles west and look for Middle Fork Lane/CR 20A. Take that for 1 mile, then turn right onto NF 4550 and drive 0.8 mile to the creek. To reach Site D at Elk Creek, which leads to Susanville from the east, drive about 14.8 miles from Deerhorn Campground on CR 20A to the west side of Elk Creek at Galena, turn right (north) onto NF 914 and go about 0.6 mile. This turn is about 20.3 miles from US 395 to the west. The remnants of the Badger Mine are about 0.8 mile up. The coordinates for Site E mark the turn for Armstrong Creek, which dries up early in the summer.

Many of the adits along Elk Creek drive straight into bedrock and reveal the quartz and calcite stringers miners sought to exploit.

Prospecting

There are numerous lode mines, prospects, and placer diggings throughout the upper section of the Middle Fork of the John Day River. Miners heavily dredged the Middle Fork back in the day, yielding considerable gold, and Vincent Creek, Granite Boulder Creek, and Elk Creek supplied much of the color.

Vincent Creek has numerous active claims, especially near these coordinates, and as of 2022 there were some club claims here as well, but the access at the Site A coordinates is open. You can drive a long way using this road, past the Morning Mine and all the way up Vinegar Hill, depending on your goals. If you do, keep an eye out for adits, tunnels, and prospects, and bring a good crack hammer. You can find a good combination of an inside bend and open, public land at Site B, the old Deerhorn Campground. Site C is on Granite Boulder Creek, past the China Diggings and other workings, and so far that section isn't claimed. Site D at Elk Creek, which leads to Susanville, hosted multiple active claims into the 1990s, but I did not see any posted signs there in 2022, and MineCache reports no active placer claims as well. There is bedrock in the creek bed at points along here, but it's a bit of a hike down. Stay out of any buildings at Susanville. Just about any creek leading into the Middle Fork from the north is worth testing. The famed Armstrong Nugget, an 80-troy-ounce behemoth (at 12 troy ounces per pound, that is nearly 7 pounds), came from Armstrong Creek. If you want to see the nugget, check at the US Bank in Baker City, at 2000 Main Street.

43 John Day

See map on page 85.
Land type: Creek, pits and tailings piles
County: Grant
GPS: Quartz Gulch: 44.37602, -118.91479
Best season: Late summer
Land manager: Malheur National Forest
Material: Fine gold, small flakes, black sands
Tools: Hammer; pan if there is water
Vehicle: Any; 4WD suggested
Special attraction: John Day museums, especially Kam Wah Chang State Heritage Site
Accommodations: Primitive camping near Canyon Mountain trailhead
Finding the sites: From John Day drive south on US 395 for 1.9 miles, then turn left onto Main Street. Follow it for 0.3 mile, and then take the slight left onto Canyon City East/Maryville Road. Drive 1.5 miles and take a right onto Gardner Ranch Road/CR 77. Follow it for 0.3 mile, then take the right turn for Canyon Mountain Trail Road. Drive 0.4 mile and stay right, going just 0.1 mile and taking the left turn up the mountain. You'll soon see the ruins of a small mining camp, and after 0.2 mile the road makes a sharp switchback to the site.

Prospecting

The John Day area has a rich mining history, with considerable dredging on the river itself, and especially near the mouths of Canyon Creek, Little Pine Creek, and at Dixie Creek in Prairie City. There is so much private land here, however, that you need to know someone, get away from the main areas and head up into the hills, or head farther downriver toward Picture Gorge. One possible exception is the new city park on the banks of the John Day River at 44.4219, -118.9569. The mouth of Canyon Creek is just west of the bridge at this park. If you give it a try, be sure to fill in your holes and leave no trace, pick up any garbage, and otherwise be a good citizen. We had high hopes for finding additional access to Canyon Creek, but other than a possible spot with a rope swing that kids use as a swimming hole, everything is out of bounds. Up the mountain near the lode mines, a lot of placer claims dot the banks of

Large chunk of quartz with rusty staining and a small streak of green malachite

Little Pine Creek. We found extensive quartz rocks in the road, and some had rusty staining and even a little green malachite.

SOUTHWEST OREGON

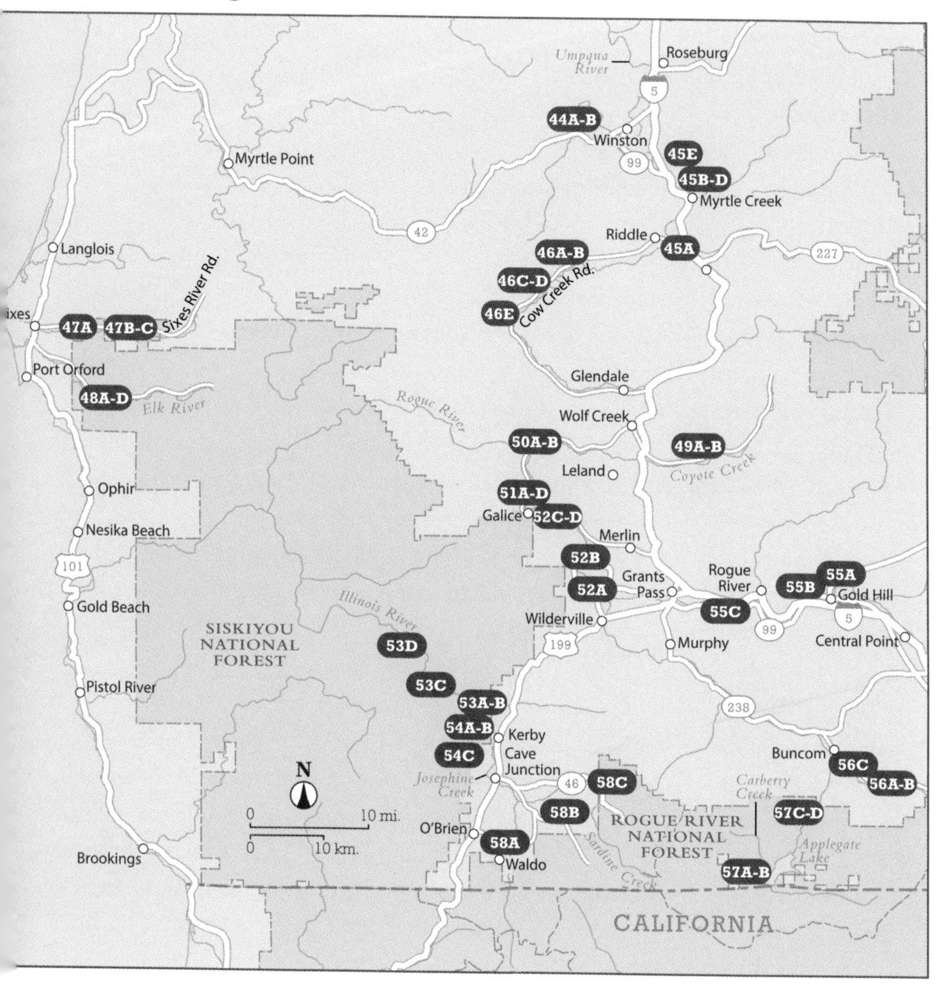

44 Lookingglass Creek

Land type: Creek
County: Douglas
GPS:
A - Wayside: 43.11778, -123.42601
B - Mouth: 43.11804, -123.42853
Best season: Any
Land manager: BLM–Medford
Material: Fine gold, small flakes, black sands
Tools: Pan, sluice
Vehicle: Any
Special attraction: Winston Wildlife Safari
Accommodations: Motels in Winston and along I-5. Twin Rivers RV Park in Roseburg is charming and right on the river.
Finding the sites: From the intersection of OR 42 and OR 99 in Winston, head west on OR 42. Just 0.7 mile from the intersection, you will approach Civil Bend on the South Fork of the Umpqua River. There is a good BLM wayside and boat ramp right on the banks of the river where you can now park (previously you could only park along the highway near the bridge). There is almost too much brush, but it is possible to make your way along the river to the mouth of the creek and then work your way up to beneath the bridge.

Prospecting

Old panning guides show a productive area on upper Lookingglass Creek where placer miners worked the gravels, but that is all private land now. This spot is open mostly for local fishing and swimming, but it *is* open, as are the banks of the South Fork of the Umpqua at the new rest area. If someone in your party needs the rest area facilities, park there, but otherwise use the coordinates for Site B to park closer to the action. There is a lot of bedrock here when the water is low, and you would do fine working to clean out the bedrock cracks. By late summer the runoff from local farms and dairy operations makes Lookingglass Creek less attractive, and you may want to work closer to the mouth, where it mixes better with the river. Still, there is plentiful black sands and good color, up to flakes, in the gravels by the rocks. I found

During low water look for traps in the bedrock and behind big rocks at the mouth of Lookingglass Creek. This locale is under the bridge on OR 42, where the creek enters the South Fork of the Umpqua River.

more colors per pan here than I did at several locations on the Coquille River below Powers, so this spot stayed in and the Coquille River ended up in the "Honorable Mentions" section at the end of the Oregon chapter.

45 South Fork of the Umpqua

See map on page 107.
Land type: River
County: Douglas
GPS:
A - Lawson Bar: 42.94828, -123.33569
B - Highway bridge: 43.02508, -123.29726
C - Western Avenue: 43.01938, -123.30014
D - Dole Road: 43.03114, -123.31104
E - Babysitter: 43.06411, -123.34481
Best season: Late summer
Land manager: BLM–Medford
Material: Fine gold, small flakes, even nuggets
Tools: Pan, sluice, highbanker, dredge
Vehicle: Any
Special attraction: Myrtle Creek
Accommodations: RV parks and motels at Myrtle Creek
Finding the sites: To reach Lawson Bar, take I-5 exit 102 and turn onto Gazley Road headed west. At the western edge of the interchange, turn left onto Lawson Bar Road. Drive just 0.7 mile and enter the recreation area. To reach the highway bridge site, take I-5 exit 108 at Myrtle Creek and head for the bridge, but do not cross it. The parking area just north of the bridge leads to the gravel bar below. To reach the Western Avenue site, resume on Main Street and go across the bridge, then take two right turns as Main Street snakes through downtown and goes south. Look for Western Avenue on your right and take it about 0.25 mile, across the railroad tracks, then take a left and work your way to the large gravel bar. The best bedrock exposures are beyond the railroad bridge. To reach the Dole Road area, also use exit 108 and take Main Street across the bridge, but instead of taking the first right turn, take the first left onto Dole Road. Follow it for 1.2 miles and start looking for parking below the road. The final spot, Babysitter, is near I-5 exit 112. After leaving the interstate, circle over to the eastern side of the highway and look for Ruckles Drive, which leads to the Rivers Edge Campground. There is a dirt track leading down to the water as soon as you start on Ruckles.

Highbanker working the old gravels at Lawson Bar, where Cow Creek meets the South Fork of the Umpqua River

Prospecting

These are all good spots to work, very similar in nature, and they are right by the interstate, so you have easy access. If a spot turns out to be blocked by construction or some other change, you can easily move on or even find another spot not listed here. Lawson Bar is a great place to start because Cow Creek enters the Umpqua system here. There is a significant spine of bedrock along the middle of the bar that you can dig into for nice gold. It is not uncommon to see a highbanker or dredge working this spot in the summer. The bridge site is easy to find and features nice bedrock as you angle to the right, upstream. At the Western Avenue site, you also need to angle back south, under and past the railroad bridge, to locate the exposed bedrock area. This spot consistently yields good color and small flakes if you are in the right spot and digging down deep. The Dole Road site is also popular and offers bedrock traps along nearly a mile of riverbank. Author Tom Bohmker (2010) calls the spot farthest north the "Babysitter" site, as one of his friends would often bring children up here and work the crevices while the kids played. There are excellent bedrock traps to the right, upstream, at the top of the bar here.

46 Cow Creek

See map on page 107.

Land type: Creek

County: Douglas

GPS:

A - Island Creek Day Use Recreation Area: 42.91325, -123.48203

B - Iron Mountain access: 42.90441, -123.53493

C - Creek access: 42.89176, -123.55819

D - Bridge: 42.86671, -123.57614

E - Panning sign: 42.77205, -123.56901

Best season: Late summer

Land manager: BLM–Roseburg

Material: Fine gold, small flakes, black sands

Tools: Pan, sluice, highbanker, dredge

Vehicle: Any, if you park safely; 4WD required if you try to get close to the creeks

Special attraction: Riddle Nickel Mine

Accommodations: BLM campgrounds at Cow Creek at 42.7722, -123.5702 and Skull Creek at 42.7722, -123.5702. Island Creek Recreation Area is day-use only.

Finding the sites: From I-5 take exit 103 and head for Riddle. If coming from the south, you could leave I-5 at Canyonville or Yokum Road. From Riddle head west on the Riddle Bypass Road, past the nickel mine, about 3 miles west of Nickel. In about 0.5 mile the Riddle Bypass Road turns into Cow Creek Road, but the turn here at the gate is a good point to zero your mileage. Drive about 4.2 miles to reach the parking spot for the Island Creek Recreation Area. The access spot for the Iron Mountain area is 8.6 miles from the nickel mine gate. There is good access to Cow Creek at 11.6 miles from the nickel mine. The bridge is 14 miles from that gate, and the sign for the panning area is near Skull Creek Coast Guard.

Prospecting

This is an excellent area for recreational panning, sluicing, highbanking, or dredging. Note that the Cretaceous bedrock in this area hosts ammonite fossils at various locales. Cow Creek has consistent gold, thanks to multiple hard-rock mines on Grayback Mountain and Diamond Butte, as well as gold and silver values coming in from Silver Butte via the Middle Fork.

The public area on Cow Creek is well marked, and there is good parking about 0.2 mile north of this sign, with easy access to bedrock along the creek.

Bohmker (2010) reports coarse nuggets way up at Galesville Reservoir, near the headwaters of the creek. He also describes the variations in the creek as it narrows, crossing different bedrock, and I highly recommend his books for anyone getting serious about prospecting in areas he covers. The Island Creek site has a lot of room to explore, multiple bedrock exposures, and plenty of boulders to move. Look for shady spots if possible—this area gets hot in the summer. If you have a good 4WD vehicle, take some time to explore and creep around on the roads until you find a good spot. There is a long gravel bar starting on the northwest part of the area, paralleling Cow Creek Road.

There is a good inside bend at the Iron Mountain access, and another one up the road at the next creek access spot at Site C. Although good access, it probably had the fewest colors in the pan, as there was less bedrock to poke at. The bridge site was the scene of a pair of bad railroad wrecks years ago, and it offers good access to an inside bend. I listed the coordinates for the panning sign, but the best parking area is about 0.2 mile north of that sign. You should find a little beach below here, if the water is low, with some bedrock and large boulders to prospect around. The best, coarsest gold is near the BLM campground, but there are many claim markers here, and the campground is not directly on the creek.

47 Sixes River

See map on page 107.
Land type: Small river
County: Curry
GPS:
A - Edson Park: 42.81544, -124.41241
B - West Edge: 42.80467, -124.31961
C - Main beach: 42.80478, -124.30599
Best season: Late summer
Land manager: BLM–Medford
Material: Fine gold, small flakes, black sands
Tools: Pan, sluice, highbanker, dredge
Vehicle: Any
Special attraction: Cape Blanco
Accommodations: Good camping at Sixes River Recreation Site; dispersed primitive camping above claimed areas east of recreation site
Finding the sites: From US 101 at Port Orford, drive north 5.3 miles to Sixes River Road. The public swimming beach associated with Edson Park and Campground is 4.2 miles up this road. The recreation site starts 10.7 miles up. We were able to connect to Elk River by going way out Sixes River Road, climbing Rusty Butte, clinging to the side of the mountain as the road devolved into a billy goat trail, and burning up our brakes coming back down the ridge to Butler Bar. Do not try it without a good navigator, updated maps, lengthy planning beforehand, and a GPS you trust.

Prospecting

The BLM set aside the Sixes River Recreation Site for recreational prospecting, and it remains popular. The rangers erected some interesting kiosk boards with information about the area's mining history as well. There are some rules here—no mining in the banks, for starters. The campsites right on the water, toward the upper part of the river, are excellent. If you are highbanking or dredging, you want to keep your nozzle in the water or very near it. There is good color here if you dig a deep hole. Use your best judgment on working with bedrock and big rocks. Before you go, do a web search for information

The Sixes River Recreation Site has ample camping and great access to the water, and it's open for recreational panning and prospecting.

about Sixes River gold mining, and look for "Distribution of Placer Gold in the Sixes River, Southwestern Oregon—A Preliminary Report," by Dr. Sam Boggs Jr. and Dr. Ewart M. Baldwin, two professors from the University of Oregon when I studied there. Their map will show you where the best concentrations are at Sixes River and where they found the coarsest gold. Over the years several placer operations worked the river below here, including the Corbin property near Pipeline Creek, the Divelbliss claim near Little Dry Creek, and the Sunrise operation west of Plum Trees. There is room to explore above the recreation site, but much of this area is claimed when gold prices are high—especially the spots with the best camping and access. There are old lode prospects up at Rusty Butte, for example.

48 Elk River

See map on page 107.
Land type: River
County: Curry
GPS:
A - Myrtle Group: 42.71165, -124.37492
B - Elk bend: 42.71711, -124.35687
C - Elk River Placer: 42.70831, -124.33191
D - Big Sunshine Placer: 42.7149, -124.3061
Best season: Late summer
Land manager: BLM–Medford
Material: Fine gold, small flakes, black sands
Tools: Pan, sluice, highbanker, dredge
Vehicle: Any; 4WD suggested
Special attraction: Oregon coast highway
Accommodations: Multiple developed campgrounds such as Butler Bar and Sunshine Bar; dispersed camping all along upper Elk Creek
Finding the sites: From Port Orford on US 101, drive north 3.1 miles to Elk River Road/NF 5325. Drive 10.4 miles to reach the old placer camp known as the Myrtle Group. The spot on the sharp bend has no camping but shows good color. It's 12.1 miles from US 101. The old Elk River Placer Mine site is 13.9 miles from US 101, and the Big Sunshine Placer is 15.9 miles from US 101.

Prospecting

Elk River has good color and plenty of black sands. This was never a major mining district, although the Bonanza Basin area on Boulder Creek, way up above Laird Lake, was said to have hosted a sizable collection of miners at one time. Your best bet now is to search for bedrock showings on the inside bends, dig a big hole at the top of a gravel bar, or scrape a little moss and clean it in a five-gallon bucket. There are a few lode claims in this valley, but the action was in placer camps such as the Myrtle Group at Site A, the Elk River Placer Mine at Site C, and the Big Sunshine Placer at the mouth of Sunshine Creek,

This area was the site of the Myrtle Group Mine, which worked the bedrock and boulders. This is a good camping spot, with plenty of shade.

above Sunshine Bar at Site D. Butler Bar has a nice campground, which is always welcome. Similarly, at the other end of the valley near US 101, the Oregon Department of Fish and Wildlife has a good fishing-access spot at 42.7861, -124.4811, but there weren't any colors to speak of.

49 Golden

See map on page 107.
Land type: Creek
County: Josephine
GPS:
A - Golden buildings: 42.68178, -123.33102
B - Robinson Gulch: 42.68372, -123.27544
Best season: Late summer
Land manager: BLM–Medford

The old church is among a handful of buildings still standing at the old ghost town of Golden.

Material: Fine gold, small flakes, black sands
Tools: Pan only
Vehicle: Any; 4WD suggested for exploring
Special attraction: Golden ghost town
Accommodations: Dispersed camping beyond Fandora
Finding the sites: From I-5 take exit 76 at Wolf Creek and angle back south and east on Coyote Creek Road. We panned a quick sample at the bridge on Miller Gulch, about 2.4 miles up the road, and found some access another 0.25 mile up the road. The ghost town is about 3.5 miles up Coyote Creek Road, and Robinson Gulch is about 3.4 miles farther up the road.

Prospecting

Coyote Creek was a major producer beginning in the 1850s, and there are prospects and lode mines throughout the upper reaches. Near the Golden church, you can easily spot the scars from major hydraulic operations, and there are tailings throughout the area. Much of the lower creek is private land, and you will not pop out onto BLM land until you get closer to Robinson Gulch. There are plenty of black sands in the creek gravels, and decent color, but this place saw heavy use, and there is not much water to work with by the end of a hot summer. Still, the townsite of Golden is worth checking out, and there are plenty of opportunities to explore the BLM roads crisscrossing the upper reaches of Coyote Creek.

To the north, Wolf Creek is worth checking out, although there are far more placer claims there to dodge. One spot worth investigating is at 42.7186, -123.2796. Like other areas of southwestern Oregon, the rocks around Golden are predominantly Cretaceous sedimentary rocks, which in some cases contain ammonite fossils.

50 Grave Creek

See map on page 107.
Land type: Creek
County: Josephine
GPS:
A - Mouth: 42.64876, -123.58427
B - Parking: 42.64817, -123.58151
Best season: Late summer
Land manager: Rogue River–Siskiyou National Forest, BLM–Medford
Material: Fine gold, small flakes, black sands
Tools: Pan only near the Rogue River; open above that for dredging and high- ["highbanking" is one word] banking
Vehicle: Any
Special attraction: Rogue River

Frank Higgins, left, and Dirk Williams, right, work the bedrock on lower Grave Creek. This spot yielded good colors in every pan.

Accommodations: Developed fee sites near Galice; dispersed camping all along lower Grave Creek

Finding the sites: The mouth of Grave Creek is 7 miles from the Galice Resort, or about 15 miles from Wolf Creek on I-5 via Lower Wolf Creek Road. You can also reach the mouth of Grave Creek from Sunny Valley via Leland Road and Lower Grave Creek Road.

Prospecting

This is an excellent creek to prospect, and the bottom around the mouth is open to the public because it's part of the Rogue River scenic area. Park safely and work your way to the creek, and then scout to the mouth of the creek on both sides of the bridge. There is excellent exposed bedrock all along this bend in the creek, and the grasses have also trapped gold against the bedrock in many places. Be cautious about valid claim markers from here on up the creek. At the junction with Lower Wolf Creek Road, you can reach I-5 via either Wolf Creek or Grave Creek. Both had multiple placer mines working their gravels, and there are still many active claims on either road. You also hit a lot more private land, so your best bet is to be content with the lower area around the mouth.

51 Galice

See map on page 107.

Land type: River

County: Josephine

GPS:

A - Galice Creek: 42.56393, -123.59687

B - Resort: 42.56847, -123.59531

C - Rocky Gulch: 42.57449, -123.59535

D - Lucky Shot Placer: 42.63173, -123.59411

Best season: Late summer

Land manager: Rogue River–Siskiyou National Forest

Material: Fine gold, small flakes, black sands

Tools: Pan only

Vehicle: Any, until you leave the pavement. Do not explore the riverbank areas without a 4WD.

Special attraction: Galice Resort

Accommodations: Multiple developed campsites along here; primitive camping outside parks and recreation areas. Galice Resort has cabins and lodging.

Finding the sites: From Merlin proceed on Merlin-Galice Road about 12 miles to Galice Resort. Work your way back south along the river to the mouth of Galice Creek. To reach Rocky Gulch, proceed on Merlin-Galice Road just 1.1 miles to the entrance to the recreation site, and again make your way south as far as you can to the mouth of Rocky Gulch. The Lucky Shot Placer worked the bedrock about 5 miles farther down the road.

Prospecting

The Galice area was a major producer in its day, and miners still work it regularly. The creek itself contains multiple claims where miners work old channel deposits. Galice Creek Road, Peavine Road, and Rocky Riffle Road all offer access to historic mining areas, but there are gates and claims everywhere. Your best bet is to hook up with a club that maintains claims up here. Barring that, you can at least pan on the many bars along this stretch, as the Rogue is a Wild and Scenic River and withdrawn from mineral entry. The mouth of Galice Creek is a great place to start; park at the lodge and plan to stop there for a

Old stone foundations near Rand Recreation Area, above the Rogue River

meal or refreshments. (They fill growlers there, for example.) Frequent jet boats are part of the attraction while sitting on the outdoor deck of the Galice Resort on a nice summer day. Use it as a headquarters for exploring the area. The Sordy Placer was just north of Galice, the Rocky Gulch Placer farther still, and the Rand Placer was at Almeda County Park. The Scandinavian-American Placer and Lucky Shot mines operated just below the boat ramp at Argo County Park. The Lucky Shot Placer offers one of your best shots at working bedrock, and when the water is low, you should get good colors in each pan.

52 Rogue River

See map on page 107.
Land type: River
County: Josephine
GPS:
A - Whitehorse Park: 42.43364, -123.46112
B - Robertson Bridge: 42.49515, -123.48748
C - Hellgate: 42.54913, -123.52424
D - Ennis Riffle Park: 42.56316, -123.57779
Best season: Late summer
Land manager: Rogue River–Siskiyou National Forest
Material: Fine gold, small flakes, black sands
Tools: Pan and trowel only; no shovels
Vehicle: Any
Special attractions: Hellgate, Galice Resort
Accommodations: Multiple developed campgrounds along the Rogue; more primitive options past Galice
Finding the sites: From downtown Grants Pass, turn off OR 99 onto M Street, then continue onto SW Bridge Street for 1 mile. Jog left onto Lincoln and then find OR 260/Lower River Road, headed west. Follow it for 5.8 miles until you reach the Whitehorse County Park river access. To reach Robertson Bridge, continue on OR 260 west for 5.1 miles. You can also reach Robertson Bridge from Merlin, via Robertson Bridge Road. To access Hellgate, drive west from Merlin on Merlin-Galice Road for 6.2 miles to the big turnout. Ennis Riffle is about 3.6 miles farther west from the Hellgate turnout.

Prospecting

The Whitehorse site is not very busy, and there is a good beach a little west of the main swimming area. Farther west, out Gunnell Road, nearby Matson County Park offers a great inside bend with a long, open beach. On the other side of the river, via OR 260, Griffin Park has a nice beach area and is just above the old Flanagan Mine. At Robertson Bridge look for bedrock at the upstream edge of the boat ramp area and dig into crevices and cracks. Look for spots where boulders have piled up on bedrock. If you can move the rocks,

The Rogue River offers access to good bedrock showings and plenty of gravel bars and beaches. The view is near Hellgate.

the sands trapped under them should have plenty of color. Otherwise, go in the other direction and investigate the top of the bar, where heavies first start to drop during flood stage. Farther downstream, the Hog Creek boat ramp provides access to the Hellgate Placer, just downstream where the bedrock pokes out. There are good bedrock showings at the Hellgate site and a nice accumulation of sandy material. We found the same combination at the bottom of Stratton Creek Road, at 42.5519, -123.5308. Indian Mary did not offer much good panning, but Ennis Riffle Park supplied good colors among the boulders near the boat ramp. Remember, this is a Wild and Scenic River, so pan only, fill in your holes, do not be a nuisance, pick up litter, and be a good citizen.

53 Illinois River

See map on page 107.

Land type: Riverbank
County: Josephine
GPS:
A - Green Bridge: 42.24694, -123.69192
B - Deer Creek: 42.26902, -123.68714
C - Sixmile Creek: 42.29669, -123.73086
D - Briggs Creek CG: 42.37573, -123.80775
Best season: Late summer
Land manager: BLM–Medford
Material: Fine gold, small flakes, black sands
Tools: Pan only along the Illinois River
Vehicle: 4WD required except at Sixmile
Special attraction: McCaleb Ranch
Accommodations: Numerous developed campgrounds along the Illinois River, including at Briggs Creek; lots of dispersed camping opportunities as well
Finding the sites: To reach the upper sites, drive south from Grants Pass on US 199 about 23 miles to 8 Dollar Road and head west. The Green Bridge is about 2.8 miles from the highway. Instead of crossing the bridge as though you were heading to Josephine Creek, continue along the rough gravel road to access a great gravel bar forming on the inside bend of the river. To reach Deer Creek, you will need a good 4WD vehicle. Backtrack on 8 Dollar Road about 0.25 mile to NF 016 and turn left, then make sure everything loose in the car is put away. The Anderson Placer at the mouth of Deer Creek is about 2.1 miles ahead on this very bumpy road. To reach the old placer mine at Sixmile Creek, drive about 20 miles south of Grants Pass on US 199 to Illinois River Road and turn west. After about 3.5 miles you will reach the trailhead to Kerby Flats, which also gives you access to Deer Creek. Sixmile Creek is about 4.3 miles farther down the road, and there is good parking there. The Briggs Creek trailhead and campground is 18.3 miles out on Illinois River Road, so another 10.5 miles of increasingly rough road.

The bridge at the Briggs Creek trailhead leads to a large gravel bar forming where Briggs Creek dumps into the Illinois River.

Prospecting

There are some great spots to check out all along the Illinois River, and during the summer months the water is inviting and the sun lights up the beaches. There were numerous placer mines throughout this area, especially just past the Green Bridge below Josephine Creek. Heading downstream, there was the Anderson Placer, the Revell Placer, and diggings at Sixmile Creek, Black Bear, Sherman, and finally Panther Creek. Obviously, there are new rules nowadays, and you will not move a lot of material like the old-timers did. Bring your biggest pans and screens and look for crevices in bedrock to clean out. This is a Wild and Scenic River, so fill in all your holes, and pick up any trash someone else left behind. Look for tiny, silvery chunks of josephinite, and save your black sands.

54 Josephine Creek

See map on page 107.
Land type: Creek
County: Josephine
GPS:
A - Rest area: 42.24246, -123.68621
B - Ford: 42.22054, -123.70774
C - Silver Nugget: 42.18691, -123.71403
Best season: Late summer for low water
Land manager: Rogue River–Siskiyou National Forest
Material: Fine gold, small flakes, black sands; josephinite, garnets
Tools: Pan only at rest area; upper zones open for all devices
Vehicle: Any; 4WD suggested for rest area, required for other sites
Special attraction: Illinois Wild and Scenic River
Accommodations: Developed and dispersed camping along Illinois River
Finding the sites: To reach the rest area site, go south from Grants Pass on US 199 about 23 miles to 8 Dollar Road/NF 4201 and turn right. Proceed about 3.2 miles as the road crosses the Illinois River and climbs up the hill; the rest area is on your left. To reach the ford, drive another 0.6 mile to NF 029, stay left, and go about 1.9 miles to the water. Theoretically, this will take you to the upper Josephine area, but it's best reached via Kerby, Oregon, which is about 26 miles south of Grants Pass on US 199. To reach the upper parts of Josephine Creek from Kerby, take Finch Road west about 0.8 mile, across the Illinois River, and turn right onto Westside Road. Go about 0.5 mile, turn left onto NF 011, and then drive 2 miles to Tennessee Pass. Take the sharp left turn and drive 2.2 miles to the water. It is a nasty road. To reach the old Silver Nugget claim from Tennessee Pass, take the "middle" option and drive 1.3 miles. If you go right, you just end up on top of the mountain. The old miner's trail will actually take you from Tennessee Pass to the rest of lower Josephine Creek with the right rig, but it looks more like a burro trail on the maps.

Prospecting

Josephine Creek contains not only gold but also josephinite, a rare silvery nickel-iron alloy also called awaruite. Most nickel-iron combinations come from meteorites, but not here. The upper stretches of Josephine Creek host

You can cross Josephine Creek at the Ford locale if you have the right vehicle, but I saw a lot of paint on some of these boulders.

the small silvery pebbles that erode from serpentinized peridotites and ophiolites, and there are chromite prospects out here as well. Save anything shiny, and inspect your black sands closely. You should recover garnets as well. The Silver Nugget spot at Site C offers excellent prospects for josephinite and decent color, but the gold improves as you move down the creek. The creek bed south of the rest area has been claimed in the past, including status as a GPAA claim at one time, so be on the lookout for claim markers. The bottom 0.25 mile of Josephine Creek Falls is inside the Illinois Wild and Scenic River and is open to panning only. Site B at the ford is also on a claim, but I included it for reference if you want to cross the creek to reach areas where few tourists ever go. Please note that beyond the rest area, these roads are notoriously rough and rutted and require high-clearance 4WD vehicles. During times of drought or near the end of summer, you can cross Josephine Creek and explore more of this area.

55 Gold Hill

See map on page 107.
Land type: Creek
County: Jackson
GPS:
A - Gold Nugget Recreation Area: 42.45685, -123.02976
B - Sardine Creek: 42.43611, -123.0785
C - Savage Rapids: 42.4191, -123.2281
Best season: Late summer for low water
Land manager: BLM–Medford
Material: Fine gold, small flakes, black sands
Tools: Pan, sluice, highbanker, dredge
Vehicle: Any
Special attractions: Beeman-Martin House, home of the Gold Hill Historical Society's museum, located at 504 First Avenue in Gold Hill
Accommodations: Valley of the Rogue State Park; limited dispersed camping due to population centers
Finding the sites: To reach the Gold Nugget Recreation Area from I-5, take exit 40 at Gold Hill, head north 0.3 mile to OR 99, and turn left. Go just 0.4 mile, then turn right onto Dardanelles Street, drive a short distance, and turn right onto OR 234. Drive along the river about 2.4 miles to the main parking area for the recreation area. To reach the mouth of Sardine Creek, return to Gold Hill and proceed west on OR 99. After you leave Gold Hill, the first bridge is over Sardine Creek—look for parking just past the bridge on the left side of the road. There is a steep trail down to the water's edge. To reach Savage Rapids, leave I-5 at exit 48, cross the Rogue River on Depot Street and connect to OR 99, then go 3.3 miles to a parking area.

Prospecting

At the Gold Nugget Recreation Area at Site A, stay away from the downstream portion of the park and concentrate on the far upper end, near the rapids once known as Dillon Falls. There is excellent bedrock here, and although you would expect this area to be exhausted, most prospectors stick to the central beach areas and do not explore enough of the upstream property. The last time we visited, we noticed prospectors on the south bank of the Rogue and

This small gravel bar at the mouth of Sardine Creek yields good color. This is where the creek meets the Rogue River near Gold Hill.

went to explore, only to find a BLM access spot (day use only) at coordinates 42.4564, -123.0269. At Site B at the mouth of Sardine Creek, which was a major producer long ago, you will find a small beach that has built up as the creek empties its load into the Rogue River. It's too bad the culvert under OR 99 doesn't have sharper riffles! The GPAA has long maintained the Sunshine Suzi claim on Sams Creek, northeast of Gold Hill, and it's worth checking if you're in that club. The hills on both sides of this valley are peppered with numerous prospects and lode mines, but many of the roads are gated, which hampers exploration. The Savage Rapids locale at Site C was once underwater behind Savage Rapids Dam, but engineers demolished the structure in 2009. This spot is gaining popularity after noted southwest Oregon author and gold-mining historian Kerby Jackson produced a YouTube video about the spot. There is limited room here, but lots of bedrock and a large boulder field to explore. Note that there are numerous county parks and access points on both sides of the Rogue in this stretch, including Seaman Bar, Coyote Evans, Pierce Riffle, Chinook, Tom Pearce, and Lathrop.

56 Little Applegate

See map on page 107.

Land type: Creek

County: Jackson

GPS:

A - Tunnel Ridge Recreation Area: 42.15831, -122.90338

B - Little Applegate Recreation Area: 42.15061, -122.87777

C - Buncom: 42.17433, -122.99712

Best season: Spring–fall

Land manager: BLM–Medford

Material: Fine gold, small flakes, black sands

Very little remains at Buncom, at the bottom of Sterling Creek on Little Applegate Road. Most of the surrounding land is private, but farther up, the Little Applegate is open to the public.

Tools: Pan, sluice
Vehicle: Any
Special attraction: Jacksonville
Accommodations: Clear Creek CG
Finding the sites: To reach Tunnel Ridge Recreation Area from Jacksonville, take OR 238 about 7.6 miles south to Ruch and turn left onto Upper Applegate Road. Drive 2.8 miles to Buncom and explore there, then take Little Applegate Road and drive 9.7 miles. You could also come in from Jacksonville via Sterling Creek Road and pop out at Buncom. Little Applegate Recreation Area is about 1.9 miles east of Tunnel Ridge. You can also reach the Little Applegate site via I-5 at Talent by taking Rapp Road, Wagner Creek Road, and then Anderson Creek Road.

Prospecting

The Tunnel Ridge and Little Applegate Recreation Areas have been open to the public for years, but they still produce. The gold is good all along the Little Applegate, with plenty of water, bedrock to inspect for cracks and traps, and large boulders to move around. You should get colors in every pan if you are patient with your prospecting—all the usual advice about going deep on inside bends, on bedrock, and under boulders applies. Be on the lookout for ticks, rattlesnakes, and poison oak, by the way. The Sterling Creek area is mostly on private land now, which is a pity, as it was a solid producer for many years. The gold coming out of Sterling Creek had a higher percentage of silver than is normal, hence the name, but it was good enough to warrant construction of a large-scale ditch system to deliver water for hydraulicking. Today there are multiple hiking trails exploring the ditch system. The coordinates for Site C at Buncom are included mostly for history and photography.

57 Carberry Creek

See map on page 107.
Land type: Creek, river
County: Jackson
GPS:
A - Carberry Creek: 42.05401, -123.16345
B - Carberry CG: 42.02527, -123.16545
C - Jackson CG: 42.11403, -123.08707
D - Gin Lin Trail: 42.11631, -123.08776
Best season: Late summer
Land manager: Rogue River–Siskiyou National Forest
Material: Fine gold, small flakes, black sands
Tools: Pan, sluice, shovel, bucket
Vehicle: Any; 4WD suggested for exploring, though roads are generally good.
Special attractions: Applegate Reservoir, Gin Lin Mining Trail
Accommodations: Carberry CG your best bet; primitive camping on Carberry Creek and along Steves Fork
Finding the sites: From just southwest of Applegate, leave OR 238 and drive south on Thompson Creek Road for 11.8 miles. The road becomes Star Gulch Road for about 0.5 mile and then becomes Carberry Creek Road, but you will barely notice. Drive another 5.8 miles, or 18.1 miles total from Applegate, and look for creek access. The campground is another 2.8 miles south on Carberry Creek Road. To reach Jackson Campground from Carberry Campground, drive another 0.4 mile east, then turn left onto Applegate Road (or Upper Applegate Road, depending on the signs). Follow the west bank of the reservoir past the dam, past Palmer Creek, for a total of 9 miles to the sign. You could also reach Jackson Campground from Ruch by driving 9.8 miles south on Upper Applegate Road. The Gin Lin Trail is 8.8 miles south of Ruch via Upper Applegate Road, then right (west) on Palmer Creek Road for about 0.9 mile.

Prospecting

This is a solid area for prospecting, but it became blanketed by claims when gold spiked in 2011, and you still need to keep an eye out as of 2022. The area was wide open in the 1990s, but I drove out Steves Fork for quite a distance in 2014 and did not find much open ground. There used to be a US

Modern 4-inch Keen suction dredge, ready to work a deep hole on Carberry Creek

Forest Service public area called Applegate #6 on Carberry Creek at 42.0492, -123.2725, but they discontinued that program. Another public panning area below the dam at 42.0663, -123.1085 closed, as did a slew of sites that are actually in California, south of the reservoir. We tried to access the Steamboat Mine at 42.0814, -123.1969 but it was gated, and it looked like all of the old townsite of Steamboat was on private land. There once was a GPAA claim on Palmer Creek at 42.1177, -123.1375, so check there if you are a member; you could also scout up there for additional land that has opened up. The Applegate area was rich in gold back in the day, especially from McKee Bridge south to the dam, but this is mostly private land now. There was so much gold at the bottom of the Applegate River that they stopped construction on the dam briefly to recover as much as they could.

You can easily reach the Applegate at Jackson Campground, and from there to about 0.4 mile south. These look like local fishing access sites mostly, both along Applegate Road on the east and via Palmer Creek Road on the west. You should be fine with hand tools along the Applegate, but if you want to run power equipment, check in at the Star Ranger Station first. Be sure to visit the Gin Lin Mining Trail and note the power of hydraulic mining, but also the incredible patience and efficiency of Chinese miners, who hand-stacked rocks as they cleared the ground. The Chinese did not just settle for the biggest, coarsest nuggets, and were able to make wages by consistently recovering fine gold.

58 Waldo

See map on page 107.
Land type: Creek
County: Josephine
GPS:
A - Waldo: 42.06086, -123.64817
B - Althouse Creek/Tartar Gulch: 42.10975, -123.52505
C - Sucker Creek: 42.13982, -123.45726
Best season: Spring–fall
Land manager: BLM–Medford, Rogue River–Siskiyou National Forest
Material: Fine gold, small flakes, black sands
Tools: Pan, sluice, highbanker, dredge
Vehicle: Any; 4WD suggested, some rough spots on gravel roads
Special attraction: Oregon Caves National Monument
Accommodations: Grayback CG; primitive camping along Althouse and Sucker Creeks
Finding the sites: To reach Site A at Waldo, drive south on US 199 from Cave Junction about 0.8 mile and turn left onto Rockydale Road. Drive 6.6 miles south and turn right onto Waldo Road, then go about 1.2 miles to reach the historical marker. To reach the Althouse Creek locale from Waldo, drive east on Waldo Road 2 miles, then turn left (north) onto Takilma Road and drive 4.6 miles to Holland Loop Road. You could also get to the Takilma Road/Holland Loop Road connection via Cave Junction by driving east on OR 46 about 1.8 miles, then turning right on Holland Loop Road for 1.9 miles. However you get here, drive 3.4 miles on Holland Loop Road, then turn right onto Althouse Road and consider stopping at the historic Holland general store for last-minute supplies. There is a turn down to the creek after about 2.1 miles. Another spot on Althouse Creek is about 0.6 mile short of the turn for the ghost town of Browntown; nothing much remains there, and it is under claim. To reach the Sucker Creek site, drive a total of 11.7 miles on OR 46 from Cave Junction and look for a picnic area/swimming hole along the water where Sucker Creek empties into Grayback Creek.

Prospecting

Waldo was a major mining district from the 1870s to the 1930s and again after World War II. It was one of Oregon's original rushes—as one of the interpretive

You should be able to pan some colors at the swimming hole where Grayback Creek meets Sucker Creek, right off OR 46 leading to Oregon Caves.

signs indicates. Sailor Diggings got its name because passing oceangoing ships could not keep a full crew thanks to the lure of riches in the goldfields. Thousands of acres were hydraulicked and dredged, especially after the miners constructed ditches to bring in water. Esterly Lakes is one result of that effort. You can still scratch around at Waldo—it looks like a giant ATV track today, and it is usually bone dry, but a lot of it is open BLM land. In 2022 there was some rehabilitation going on, so avoid that. There is scattered, limited access to the East Fork of the Illinois River, and I did spy a fishing spot along Takilma Road at 42.0864, -123.6173. Both Althouse Creek and Sucker Creek were major producers, and there are miles of tailings along both creeks. Miners dredged Althouse Creek repeatedly in the early days, and it remains pretty torn up at Tartar Gulch. There is primitive camping at Site B and good access. Beware of new claim markers, but the two spots listed here should be open. There is not much to see at Browntown, and there was an active claim there when we visited in 2014. Note that much of Sucker Creek was claimed up or gated off while gold prices were high, almost to the California state line. The spot at the bottom listed here is nice and shady, at least; Grayback Campground offers more access to water. The community swimming spot and rope swing at Site C, where Grayback Creek and Sucker Creek meet, is one of my favorite locales.

CASCADES (OREGON)

59 Sharps Creek

Land type: Creek
County: Lane
GPS:
A - Sharps Creek Recreation Area: 43.66527, -122.80695
B - Weyerhaeuser #1: 43.63801, -122.78599
C - Weyerhaeuser #2: 43.63509, -122.78395
D - Weyerhaeuser #3: 43.63338, -122.77894
E - Mineral CG: 43.58318, -122.71369

Cascades South

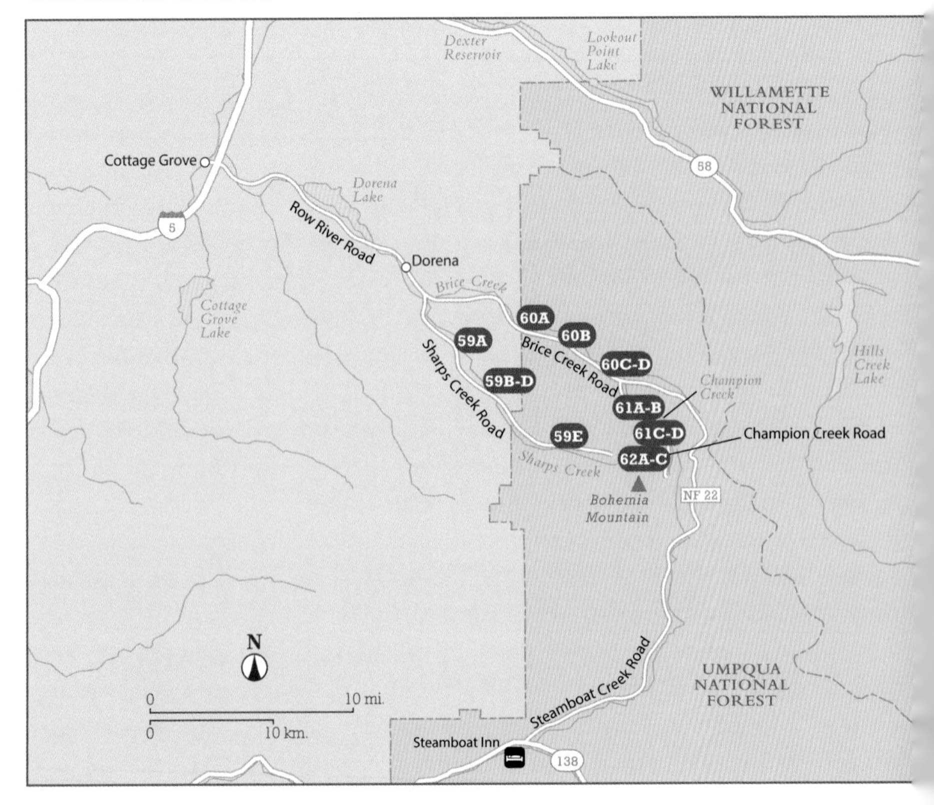

Sharps Creek has long stretches of exposed bedrock to search for traps and crevices. Try moving large boulders and developing a deep hole (which you should fill in when you finish).

Best season: Late summer for low water
Land manager: Umpqua National Forest
Material: Fine gold, small flakes, black sands
Tools: Pan, sluice, highbanker, dredge
Vehicle: Any until you get to Mineral CG; 4WD required above Mineral
Special attraction: Bohemia City
Accommodations: Sharps Creek Recreation Area is a fee campground; Mineral is free. Lots of primitive camping along Sharps Creek on USFS land, especially near Mineral.
Finding the sites: From I-5 take exit 174 at Cottage Grove and head east on Row River Road toward Dorena Lake. Go 15.6 miles as Row River Road briefly becomes Government Road and Shoreview Drive; it resumes as Row River Road quickly. At Culp Creek turn right onto Sharps Creek Road and cross the bridge. (There are numerous local-access fishing spots and swimming holes near this bridge, on both sides and in both directions, and you could get a good test pan below the bridge among the bedrock.) Set your mileage to zero at this point. Proceed on Sharps Creek Road for 3.2 miles to reach the Sharps Creek Recreation Area, which attracts many campers throughout the summer. It is open for panning.

Weyerhaeuser #1 is 2.4 miles farther up the road, and #2 is another 0.2 mile. Weyerhaeuser #3 is 0.3 mile farther, and then you are at the end of the recreation area, so please avoid the nice spots just a little farther up. At 10.1 miles from the bridge, turn left onto BLM Road 2460. (If you were to miss this turn, you would end up at the North Fork of the Umpqua River via Rock Creek Road.) Head east, uphill, on BLM Road 2460 for 12.2 miles to reach Mineral Campground. This road is already rough and is the base of the Hardscrabble Grade, which is unsuitable for sedans and minivans. Hardscrabble Grade will connect you to Bohemia Saddle, Bohemia City, Champion Creek, and eventually Brice Creek.

Prospecting

Sharps Creek contains good gold, with plenty of bedrock to search and good accumulations of flour gold and black sands. The Sharps Creek Recreation Area and the Weyerhaeuser sites all provide excellent access to Sharps Creek, but the higher up the creek you go, the fewer people you'll see. There used to be signage at the Weyerhaeuser site, and it still has a general marker on Google Earth, but nowadays the local custom seems to treat this general area as open. Your best bet is to track your mileage and play by the rules. Umpqua National Forest land starts just before the Martin Creek Bridge, which has color, and there are valid claims leading up to the Mineral Campground that you should avoid. Sailor Gulch historically contained good gold, as did nearby Fairview Creek and Puddin Rock Creek (which leads to the Star Mine). Plan to move big boulders and search for bedrock cracks that trap gold during the spring runoff. You can access Bohemia via the road to Mineral Campground, and there are multiple mines along the way.

60 Brice Creek

See map on page 138.
Land type: Creek
County: Lane
GPS:
A - USFS boundary: 43.67673, -122.74429
B - Cedar Creek CG: 43.67035, -122.70654
C - Lund Park CG #10: 43.64797, -122.67329
D - Hobo CG: 43.64642, -122.66962
Best season: Late summer for low water
Land manager: Umpqua National Forest
Material: Fine gold, small flakes, black sands
Tools: Pan, sluice, highbanker, dredge
Vehicle: Any; 4WD suggested for exploring but not necessary
Special attraction: Dorena Lake
Accommodations: Cedar Creek and Lund Park are fee campgrounds; Hobo Camp is free. Multiple primitive camping spots along Brice Creek.
Finding the sites: From I-5 take exit 174 at Cottage Grove and head east on Row River Road toward Dorena Reservoir. Drive about 18.7 miles on Row River Road, which changes names occasionally to Government Road and Shoreview Drive but eventually connects itself back to Row River Road. This main road becomes Brice Creek Road, and you proceed through Disston and drive a total of 2.5 miles to the USFS boundary. From here Cedar Creek Campground is 2.2 miles up Brice Creek Road. The entrance to Lund Park Campground is 4.4 miles from the USFS boundary; spot #10 is by far the best place to camp here. The Hobo Campground is 5 miles from the USFS boundary. The turn for Champion Creek is 5.7 miles from the USFS boundary.

Prospecting

Once you reach US Forest Service land, you can start to look for creek access and begin prospecting, but be aware that the first section of land does include a single claim, the Gold Meadow Plains claim. There are dozens of places to reach the creek, which has good gravels, large boulders, and occasional bedrock. The areas around Cedar Creek Campground, Crawfish Trailhead,

This stretch of Brice Creek at Lund Park Campground has plenty of bedrock and many boulders, and each pan yielded good colors.

Lund Park Campground, and Hobo Camp are generally reserved from locating mineral claims and thus open to the public in perpetuity; all are open for panning, sluicing, highbanking, and dredging. If a new claim marker springs up, just dodge it and move on. Once past Hobo Campground there are many more claims, so note that and react accordingly. Be sure your activities do not interfere with other recreation activities; if a family has occupied a swimming hole, you probably want to chat first before moving in with a 4-inch dredge, and that assumes you have all your permits in order. Brice Creek has plentiful black sands, lots of very fine gold, and occasional flakes and pickers if you can reach a good bedrock trap during low water or operate some kind of machinery. This is one of the major drainages for the Bohemia District, and the gold gets coarser the farther up you go.

61 Champion Creek

See map on page 138.
Land type: Creek, lode mine
County: Lane
GPS:
A - Public area: 43.62784, -122.65875
B - Adit: 43.61111, -122.63726
C - Mill ruins: 43.58272, -122.63461
D - Lower portal: 43.58086, -122.63375
Best season: June–Oct
Land manager: Umpqua National Forest
Material: Placer gold along creek; pyrites and lode gold at adits and near portal and mill
Tools: Placer tools along creek; heavy hammer at mines
Vehicle: Any; 4WD suggested to reach Champion Mine lower portal, as road is steep and rough the higher you go.
Special attractions: Smith Falls, Bohemia City
Accommodations: Lund Park CG (fee) and Hobo Camp (no fee) along Brice Creek; primitive camping all along Champion Creek. No camping at mine.
Finding the sites: From I-5 take exit 174 at Cottage Grove and head east on Row River Road about 18.7 miles. The main road is alternately called Government Road and Shoreview Drive, but it trends generally southeast past Dorena Reservoir and Culp Creek and reaches Brice Creek Road just before Disston. Drive 8.4 miles on this main road, eventually known as NF 22, following the water, until you reach a right turn for Champion Creek Road. The public area is about 1 mile up from the turn; look for the yellow rectangle sign that says "Withdrawn from Mineral Location." There are a couple of adits along this road in the cliffs, starting at about 1.7 miles farther south from the public area or 2.7 miles from Brice Creek Road. These are probably the Trixie Mine. At 5.4 miles from the turn, you can park at the reclamation area and reach the Champion Mine and mill.

Prospecting

There are GPAA claims at the bottom of Champion Creek, named the Golden Cat claim, and about 3 miles up Champion Creek Road is the Little

You will find excellent camping at the public area on Champion Creek, with a giant fire ring and easy access to the creek.

Paradise claim, but unless you are a member, you do not want to do anything more than pan a couple sample pans on them; stick to the public area at Site A. There is plenty of bedrock to explore here and good values, but the current is fast and the bedrock is smooth, so look for cracks and crevices. Champion Creek drains a major portion of the main mineralized area in the Bohemia District, so the potential is there. The road gets worse the farther up you drive, and soon becomes unsuitable for sedans and minivans. The adits are interesting because you can see exactly what the miners were exploiting. This is a rhyolite tuff with pyrite, galena, and sphalerite values in addition to gold. All the way up at the reclamation area that marks the old Champion Mine, you used to be able to find representative rock samples on the ground close to the lower portal and over by the mill. They are scarce now, but with a little scratching around, you should be able to find something. If you have a sturdy rig with solid suspension and good tires, continue up this road to the Champion Saddle and Bohemia City, and loop back to Cottage Grove via Sharps Creek.

62 Bohemia City

See map on page 138.
Land type: Ghost town, abandoned mine
County: Lane
GPS:
A - Turn to Bohemia City: 43.58082, -122.64337
B - Musick Mine main portal: 43.57929, -122.65384
C - Upper portal: 43.57809, -122.65363
Best season: Late spring–Sept
Land manager: BLM–Salem, Umpqua National Forest
Material: Sulfides (galena, chalcopyrite, sphalerite, pyrite)
Tools: Heavy hammer
Vehicle: Any; 4WD suggested
Special attractions: Fairview Lookout tower, Bohemia City
Accommodations: Multiple developed campgrounds on Brice Creek and Sharps Creek; Mineral CG on Sharps Creek is a favorite and has a restroom. Primitive camping throughout this area.
Finding the sites: From I-5 take exit 174 at Cottage Grove and head east on Row River Road toward Dorena Lake. Go 15.6 miles as Row River Road briefly becomes Government Road and Shoreview Drive; it resumes as Row River Road quickly. At Culp Creek turn right onto Sharps Creek Road and cross the bridge. (There are numerous local-access fishing spots and swimming holes near this bridge, on both sides and in both directions, and you could get a good test pan below the bridge among the bedrock.) Set your mileage to zero at this point. Proceed on Sharps Creek Road for 3.2 miles to reach the Sharps Creek Recreation Area, which attracts many campers throughout the summer. It is open for panning. From Sharps Creek Recreation Area, continue on Sharps Creek Road/BLM Road 2460, and stay on 2460 for 14.2 miles, past Mineral Campground and up Hardscrabble Grade. You will pass numerous adits in the cliffs, starting at 43.5825, -122.6976, and some deteriorating buildings. At the Bohemia Saddle, a left turn goes up the mountain to the Fairview Lookout, and a hard right leads to a picnic area. Otherwise, proceed east, down the hill about 0.7 mile, to the hard right turn that leads to Bohemia City. It is a rough road that is not suitable for sedans and minivans. Drive about 0.5 mile and park in the shade, then explore the dumps. I do not recommend going inside the dilapidated old store or any of these adits.

The view from the Musick Mine main portal shows the roof of the wrecked rail system, extensive tailings, the old powder magazine, and the general store in the distance.

Prospecting

For rockhounds, there is a nearly exhausted tourmaline deposit in the road cut near Site A where you turn to Bohemia, and there is also jasper scattered throughout this area. Once at the ghost town and the mine, look for purple-tinted rhyolite tuff impregnated with sulfides such as pyrite. Split open likely rocks, and understand that the farther from the main area you go, the more likely you are to get good samples. Under a powerful hand lens, look for small flecks of gold among the brassy pyrite. This is the Musick Mine, and the adit has a gate. There are 7,200 feet of drifts and crosscuts under you, with multiple stopes, raises, and winzes. The mine started in 1891 thanks to James C. Musick, and there are numerous other outcrops all over this side of the mountain. There are reports of decent quartz crystals from dumps above the main mine. The jeep trail heading down and east leads to more dumps and collapsed adits. Be respectful of all the remaining structures, and beware of rusty nails, rotting wood, and the other usual suspects at old mines.

63 Quartzville Creek

Land type: Creek
County: Linn
GPS:
A - Northside County Park: 44.40157, -122.73381
B - Lower bend: 44.54916, -122.42396
C - Yellowbottom: 44.58671, -122.36084
D - Canal Creek: 44.58744, -122.34727
Best season: Late summer
Land manager: Willamette National Forest
Material: Fine gold, small flakes, rare nuggets
Tools: Pan, sluice, highbanker, dredge
Vehicle: Any; paved roads
Special attraction: Quartzville site
Accommodations: Yellowbottom CG; primitive camping at multiple spots along creek
Finding the sites: From the intersection of OR 228 and US 20/Santiam Highway in Sweet Home, drive about 0.3 mile east on US 20 to 12th Street and turn left (north). Go about 0.3 mile north to Redwood Street and turn left. You will see Northside County Park; park here. Make your way across the field to the water, head west, and follow the decent trail to a large outcrop of bedrock, with multiple traps containing excellent colors. To reach Quartzville Creek, resume on US 20 and head east. About 5.9 miles from the intersection of OR 228 and US 20, turn left (north) onto Quartzville Drive. Pass the reservoirs and stay on this road for 18.4 miles. Look for a good access spot with some accumulated gravels below the road. The Yellowbottom site is 24.2 miles from Sweet Home and requires a little hike down. The mouth of Canal Creek is about 0.8 mile farther. If you have good maps and a GPS system, you can reach Quartzville Creek from Detroit, but the passes usually host late snow and contain frequent washouts.

Prospecting

Quartzville Creek is a prime spot in Oregon because an 11-mile stretch of the creek is set aside for recreational mining. Nobody can stake a claim here. If you are just passing through Sweet Home and do not have time to make the whole drive into the forest, try Northside County Park. There is an excellent

Cascades North

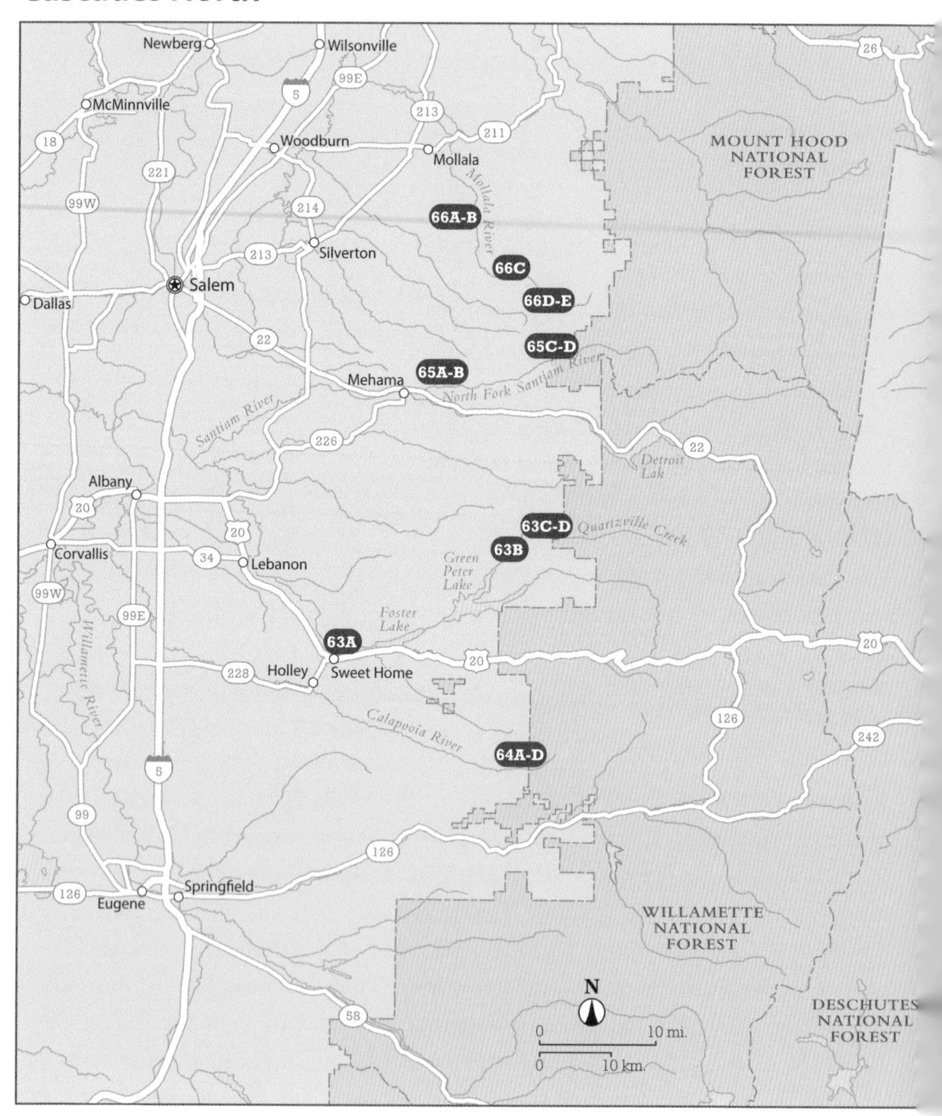

This broken bridge is on NF 11, above the recreational corridor, but it shows the power of the creek during flood stage.

exposure of bedrock just downstream that traps color and black sand continually, and it is an easy hike. The Quartzville Corridor offers reliable colors and small flakes, and will yield nuggets if you can find the right bedrock crack. You don't always need an inside bend either. We found our first nugget here working a highbanker on the "wrong" side of the creek. During the summer dredging season, you will find a lot of activity along here, so if one spot is swarming with people, just keep trying. The Canal Creek site has great bedrock, and Canal Creek supplies plenty of color, as it drains part of the Quartzville District. There are additional open areas on Canal Creek, but be aware that parts of it are sometimes claimed. If you travel east from Canal Creek on NF 11, you will find claim markers for the Star Duster up by the mangled old bridge, washed out in 1969. There have been a couple GPAA claims up here in the past as well, so if you can find an open area, try a few test pans and check it out. There are old lode mines on both sides of Quartzville Creek near the washed-out bridge. There are very few remains at Quartzville itself, and in 2014 I noticed the old historic sign was gone as well. From Canal Creek go uphill on NF 1133 for 1.7 miles and look for a gentle right turn at 44.5874, -122.3473. The town was along this stretch for about 0.2 mile to the creek washout. There are more unremarkable lode claims, abandoned and overgrown, along NF 1133 for 3.3 miles above the turn to the townsite.

64 Calapooia River

See map on page 148.
Land type: Creek
County: Linn, Lane
GPS:
A - Swim access: 44.29558, -122.65798
B - United States Creek: 44.23667, -122.38472
C - Poorman Mine: 44.224611, -122.354311
D - Crystal Hill: 44.221033, -122.349126
Best season: Late summer
Land manager: Willamette National Forest
Material: Fine gold, black sands; ore samples, quartz crystals
Tools: Pan, sluice; heavy hammer, screen
Vehicle: Any; 4WD suggested the higher you go, required at Site D
Special attraction: Brownville's Pioneer Park
Accommodations: Primitive camping beyond United States Creek; developed camping south at Blue River Reservoir
Finding the sites: From the general store near the bridge on OR 228 at Holley, Oregon, head east on Upper Calapooia Drive for 6.6 miles to the swimming area. This is Site A. To reach Site B, drive up the river another 11.8 miles, then follow NF 2820 for about 6.8 miles to the mouth of United States Creek. Note the old bridge foundations, which used to be NF 675 but washed out long ago. There are periodic issues with going through timber company lands from Holley, so you could also come in from US 20 via Canyon Creek Road. To reach the Poorman Mine and Crystal Hill from here, continue east and then south, up the hill, on NF 2820 for 6.9 miles to the saddle. Proceed another 0.6 mile, then turn right, down the hill, and follow this road for about 1.8 miles to the mine. To reach Crystal Hill, stay on NF 2820 and drive 1.6 miles from the saddle, then look for a very rough road leading up the hill. The crystal diggings are about 0.1 mile up this road. You can reach the saddle from the south, via OR 126, by turning onto Old Scout Road to reach Blue River Reservoir. Stay on this road as it becomes NF 15, about 4.8 miles from the highway, and continue onto NF 1509 for 0.3 mile, then take NF 1510 for 7.7 miles to the saddle.

This is the mouth of United States Creek, where it enters the upper Calapooia River. The best panning is slightly upstream from here, which is often claimed up, and downstream for about 3 miles.

Prospecting

The upper Calapooia River is quite scenic, and there is good rockhounding in the lower gravels for Holley blue agates and petrified wood. Unfortunately, most of the river is controlled by the local timber company. If you just want a sample from the drainage, you can check in at McKercher Park at 44.3593, -122.8764. There is access to the river and, when the water is low, some beach sand to sample. Farther up is McClun Wayside County Park, at 44.3339, -122.7551, which has similar access to limited beach sands and gravels. From here on up, things get dicey until Site A, which is a popular swimming hole. There is a slight bend in the river here and good gravels to work in. From here there is no good access to the river's edge until you reach US Forest Service land at Site B. United States Creek drains the northern part of the Gold Hill mining district, also called the Blue River District, and the mouth of the creek has excellent color. The banks of the Calapooia yield good colors from this general area, diminishing the farther away you get unless you find a good bedrock trap. Be cautious of active claims above here; they come and go as the price of gold fluctuates. There is a lot of mineralization up here, as well as many old mines, most of which have collapsed. Rockhounds have visited the Crystal Hill site for years in search of quartz and amethyst crystal scepters, but when I took mine home and cleaned them up, I found a lot of black sands and many tiny colors in the washtub. Most of my digging was from the far northern part of this area, where the land slopes off dramatically. Note that Bohmker (2010) reported decent panning below here on the south side, but Quartz Creek is difficult to reach.

65 Little North Fork

See map on page 148.
Land type: Riverbank
County: Marion
GPS:
A - North Fork County Park: 44.795588, -122.566684
B - Bear Creek: 44.800314, -122.469610
C - Salmon Falls: 44.831980, -122.371808
D - Three Pools: 44.838149, -122.308256
Best season: Late summer
Land manager: Willamette National Forest
Material: Fine gold, small flakes, black sands
Tools: Pan, sluice, highbanker, dredge
Vehicle: Any; gravel near top
Special attractions: Salmon Falls, Opal Creek
Accommodations: Shady Cove CG; dispersed camping above Three Pools; B&Bs near Elkhorn
Finding the sites: From I-5 take exit 253. Turn left (east) onto OR 22 and drive 21.9 miles to North Fork Road. After about 2.1 miles on North Fork Road, look for a right turn down to the water. Use a backpack to get your pans and tools past the heavily used public area and around to the big bend in the river. Bear Creek County Park is about 5.8 miles up the road from North Fork Park. Salmon Falls is about 6.1 miles beyond Bear Creek Park. To reach Three Pools, drive another 1.5 miles beyond Salmon Falls and, after leaving the pavement, continue onto NF 2207. Drive 1.2 miles, then look for a right turn leaving the main road to Jawbone Flats and headed down the hill; this is NF 2207. In a good year this road winds all the way to French Creek Road and Detroit Lake. In a bad year there is snow most of the year, or washouts, or boulders in the road.

Prospecting

Tom Bohmker's book *Gold Panner's Guide to the Oregon Cascades* covers this area well. He lists several more spots along the Little North Fork, where he usually got color. At the lowest spot, at North Fork County Park, you will probably want to angle around to the bend where the gravels first start

Jawbone Flats was a significant mining area at one time but is now off-limits for mining. The road to Detroit from here is only open from late June to September.

to accumulate. At Bear Creek you will find a good section of the river for prospecting, with bedrock and gravel bars appearing during low water. At Salmon Falls we hiked above the falls, where there is ample bedrock showing, and even farther up to the large bend, where there is bedrock showing by the deep pool. Up where the pavement ends, there are some big rock bluffs, and if you feel like a strenuous hike, you can angle down to the bottom of the gorge. Finally, at Three Pools you can park and walk in to the top of the rapids, where there are countless bedrock traps to work. There are numerous lode mines and prospects at the top of the drainage, including the Silver King Mine at Henline Falls and adits on Horn Creek and Tincup Creek. There are additional mines along the trail to Jawbone Flats, including the Santiam Copper Group prospects and the Mandalay Mine. This area suffered extensive forest fires in 2020, and access may not be restored until 2023.

66 Molalla River

See map on page 148.
Land type: Creek
County: Clackamas
GPS:
A - Three Bears: 45.01713, -122.48344
B - Mama Bear: 44.995784, -122.489108
C - Horse Creek: 44.962050, -122.434445
D - Bridge: 44.931052, -122.350043
E - Copper Cliffs quarry: 44.926363, -122.342623
Best season: Late summer
Land manager: Santiam State Forest
Material: Fine gold, small flakes, black sands
Tools: Pan, sluice, highbanker, dredge
Vehicle: Any
Special attraction: Shotgun Falls, near Ivors Landing
Accommodations: Cedar Grove CG before Ivors Landing; dispersed camping discouraged
Finding the sites: From the intersection of OR 211/Mollala Avenue and Main Street in Molalla, drive south on Mathias Road 0.25 mile, then take Feyrer Park Road 1.7 miles east to Dickey Prairie Road to Feyrer Park. Swing south on Dickey Prairie Road and drive 5.2 miles, staying on Dickie Prairie until you reach the bridge. Turn right, cross the river, and turn right on South Molalla Forest Road. Drive 5 miles to the Three Bears area. Mama Bear is 1.9 miles farther up. Go another 4 miles up South Molalla Road to the Horse Creek Road turnoff and drive about 0.3 mile to the bridge. To reach the upper area, return to South Molalla Road and drive 1.6 miles. Turn right onto Rooster Rock Road and go about 0.6 mile to the bridge. Look for active claim markers before committing to working hard. The Copper Cliffs quarry, about 0.6 mile up, along the creek, contains agate seams and zeolites but not much copper. The Ogle Mountain Mine, farther up this road at 44.8925, -122.3352, is on private land.

Try the Molalla River when the water is low and you can reach some of the gravelly areas in the basalt gorge.

Prospecting

The Molalla River starts way up on a ridge that separates the drainage from the Little North Fork of the Santiam River, and while the mineralization is there, it is limited. You will find mostly black sands and small gold on the Molalla, but it is a nice short drive from Portland and a good place to get in some practice. It will test your skills, but you should be fine if you can get to a good bedrock trap. If you have your permits in order, you can dredge here too. The 4-mile stretch between Ivors Landing and the Horse Creek bridge is an open recreation area for gold prospecting, but that seems like it could change any year. The Cedar Grove Campground before Ivors Landing will limit dispersed camping, so know that conditions are changing up here.

HONORABLE MENTIONS

The following Oregon locales did not make it into the book, but they're worth a visit if you're in the area.

J. Chetco River

The Chetco River is open for panning, but it is a Wild and Scenic River and you can only work below the high-water mark, so no digging in the banks, no dredging, no sluicing . . . it is very restricted. The panning is good if you find a nice trap, but recently there has been even more agitation to lock up the river completely and ban mining for the whole length. I do not recommend spending much time learning it.

K. Coquille River

There are a few places on the Coquille River where you can find bedrock traps and good black-sand accumulations. I have not had a lot of success panning here, so I am not convinced a dredge would be worthwhile. There do not seem to be many active claims on it, which is usually a sure sign that experienced prospectors are not willing to invest much time and money here.

L. Camp Carson

This old hydraulic mine along the upper Grande Ronda River, south of La Grande, is reclaimed and reverting to wilderness. The old diggings are not easy to find anymore, although there are many river rock tailings below the main pit. I panned a few colors below Camp Carson, but it was not that inspiring, and you are probably better off searching for huckleberries and morel mushrooms, observing the elk, and looking for agates near Ukiah.

M. Fall Creek

The Middle Fork of the Willamette River was never a major producer. I have found some fresh work around the Fletcher Prospect, up a decent trail at approximately 43.9181, -122.3989, but the Ironside Mine saw the most tunneling, with just 210 feet of total workings—not very impressive (Brooks and Ramp, 1968, p. 308). Billy Creek meets Christy Creek at 43.902234, -122.385441, but the lower part of Christy Creek is much more accessible.

N. Malheur

I have been out to Rye Valley, down to Amelia, and there is not much water. The China Diggings spot on Willow Creek near Brogan yields some colors, but it is not a favorite of mine.

O. Myrtle Creek

The mouth of Myrtle Creek on the South Fork of the Umpqua is good, but the upper area is mostly private land. I would rather play at Lawson Bar.

P. Pike Creek

Never that great, and the nearby Alvord Ranch has finally reached the end of its patience with tourists camping all over the ranch, leaving trash, and knocking down gates. The nearby Alvord Hot Springs is still open, for a small fee, and the ranch allows fee camping above the baths. Land status is changing rapidly out here, but back in the day, it was an interesting hike up the creek to the old uranium mines.

Q. Quartzburg

This frustrating district is located on private land on Dixie Creek, which may have given up more than 22,500 ounces of gold. Nowadays the tailings piles are disappearing, and the roads do not go through. CR 58/Dixie Creek Road crosses the creek at 44.528436, -118.702083, and I was able to get a quick sample pan. You could probably reach Dixie Meadows from the north, via the Middle Fork of the John Day, but I have not tried that yet.

R. Scissors Creek

The Ochocos hosted active claims in the recent past, so you know you can still find color here, just not much. There's an impressive adit at 44.412915, -120.364322; go up the trail about 100 yards from the pull-out, find where the trail crosses the creek, and follow it to the mine. There are additional adits all over the hills between here and the Scissorsville Historic Site sign in the meadow along the main road.

S. Spanish Gulch

The placer area at Spanish Gulch drew interest in the 1860s, centered on Rock and Birch Creeks near Antone. The quartz vein mineralization is interesting (Brooks and Ramp, 1968, p. 161), but both the lower placer area and the upper lode mines are on gated, private land and off-limits. We drove up here in 2014 and ran into a gate, and then a ranch official chased us off.

Part III: Idaho

Idaho

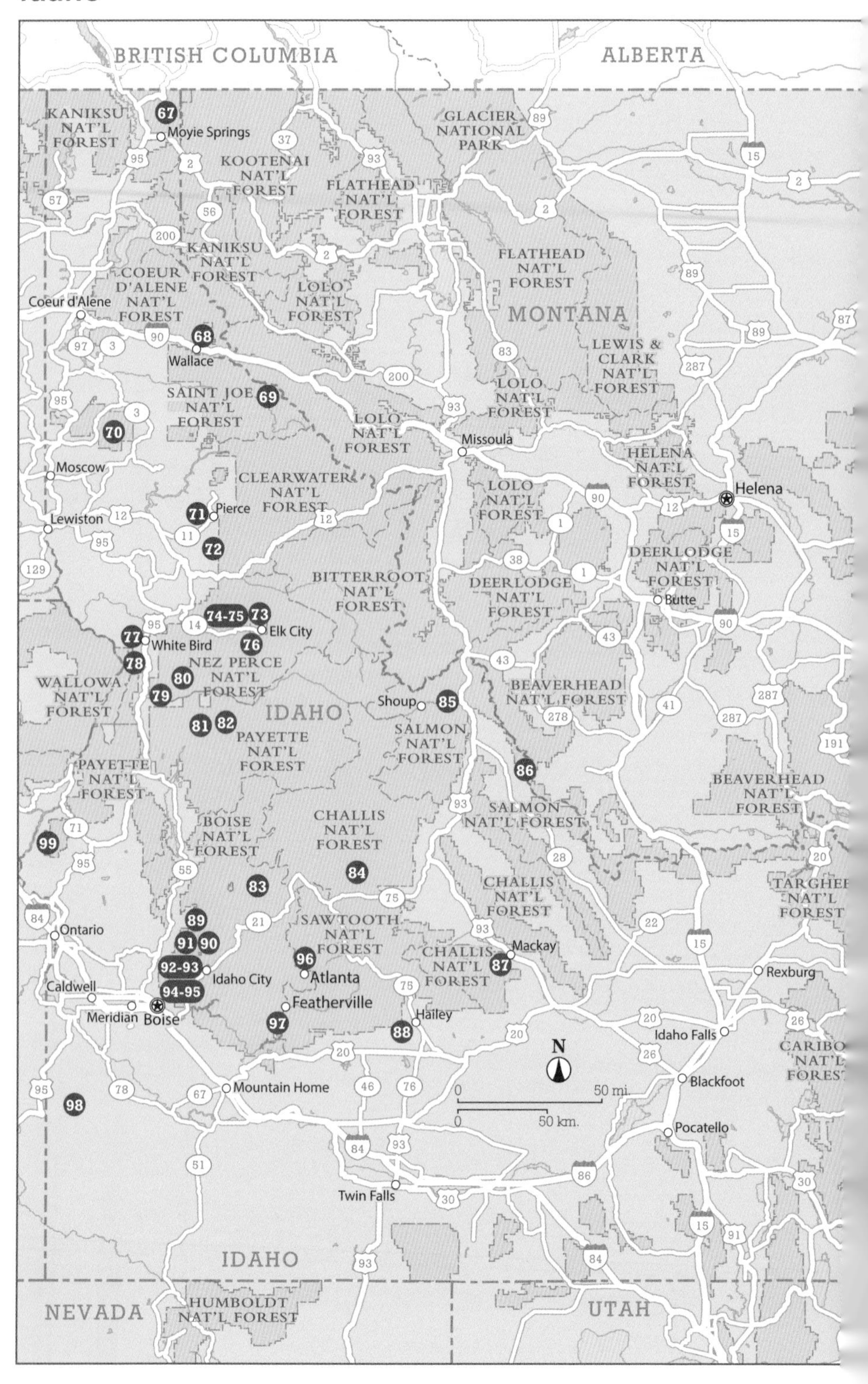

BRITISH COLUMBIA
ALBERTA
KANIKSU NAT'L FOREST
Moyie Springs
KOOTENAI NAT'L FOREST
GLACIER NATIONAL PARK
FLATHEAD NAT'L FOREST
KANIKSU NAT'L FOREST
COEUR D'ALENE NAT'L FOREST
LOLO NAT'L FOREST
FLATHEAD NAT'L FOREST
MONTANA
Coeur d'Alene
Wallace
LEWIS & CLARK NAT'L FOREST
SAINT JOE NAT'L FOREST
LOLO NAT'L FOREST
LOLO NAT'L FOREST
Missoula
Moscow
CLEARWATER NAT'L FOREST
HELENA NAT'L FOREST
Helena
Pierce
Lewiston
LOLO NAT'L FOREST
DEERLODGE NAT'L FOREST
BITTERROOT NAT'L FOREST
DEERLODGE NAT'L FOREST
Butte
Elk City
White Bird
NEZ PERCE NAT'L FOREST
WALLOWA NAT'L FOREST
BEAVERHEAD NAT'L FOREST
Shoup
IDAHO
SALMON NAT'L FOREST
PAYETTE NAT'L FOREST
PAYETTE NAT'L FOREST
BEAVERHEAD NAT'L FOREST
SALMON NAT'L FOREST
BOISE NAT'L FOREST
CHALLIS NAT'L FOREST
CHALLIS NAT'L FOREST
TARGHEE NAT'L FOREST
SAWTOOTH NAT'L FOREST
Ontario
Mackay
CHALLIS NAT'L FOREST
Idaho City
Atlanta
Rexburg
Caldwell
Featherville
Meridian
Boise
Hailey
Idaho Falls
CARIBOU NAT'L FOREST
N
Blackfoot
Mountain Home
0 50 mi.
0 50 km.
Pocatello
Twin Falls
IDAHO
NEVADA
HUMBOLDT NAT'L FOREST
UTAH

PANHANDLE REGION (IDAHO)

Idaho Panhandle

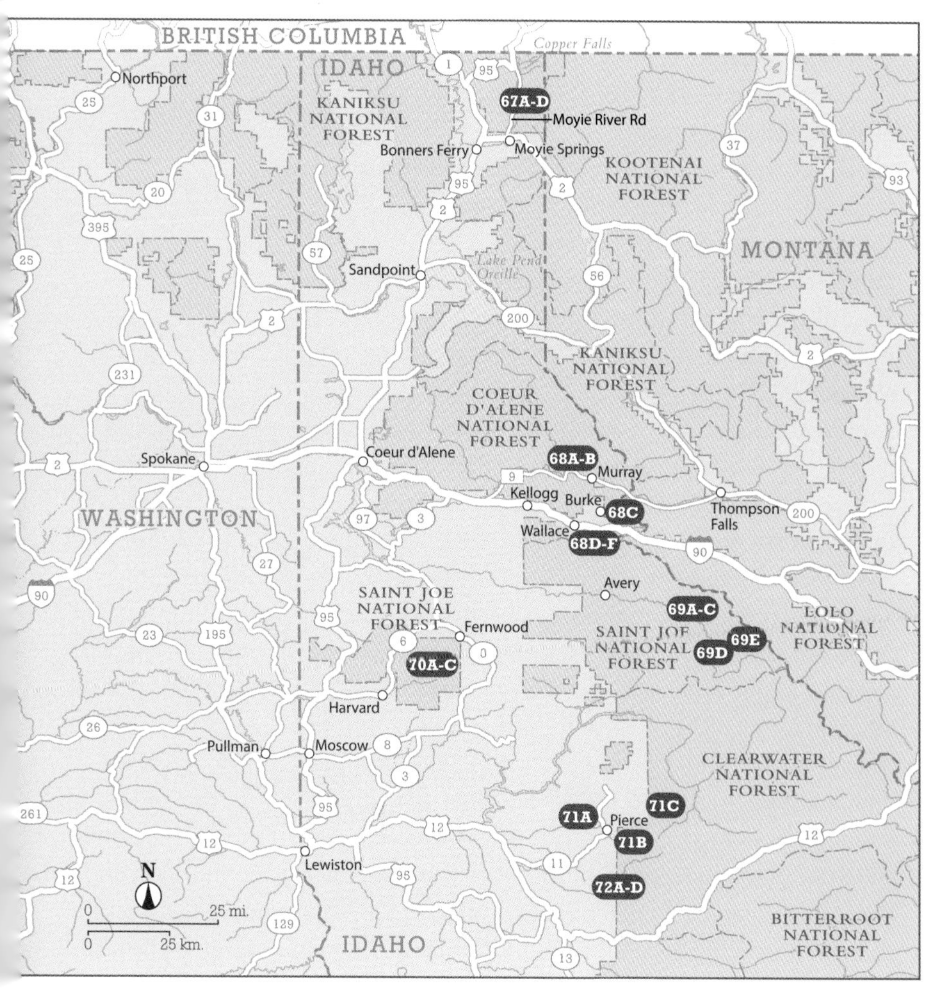

67 Moyie River

Land type: Creek, river
County: King
GPS:
A - Meadow Creek CG: 48.81969, -116.14652
B - Placer Creek: 48.83782, -116.13851
C - Bridge: 48.82495, -116.16574
D - Moyie Crossing: 48.91993, -116.17922
Best season: Late summer
Land manager: Idaho Panhandle National Forests
Material: Fine gold, small flakes, black sands; garnets
Tools: Pan, sluice, highbanker, dredge
Vehicle: Any; 4WD suggested—roads are rough in places.
Special attractions: Moyie Dam, Copper Falls
Accommodations: Copper Creek and Meadow Creek CG; dispersed camping along the Moyie River
Finding the sites: Most folks start from Moyie Falls, but do not try to take the road that closely follows the Moyie River north of town; it is extremely rough in places. Instead, take Meadow Creek Road, about 2 miles east of US 95, or 3.5 miles west of Moyie Falls. After 0.3 mile the road takes a hard right to stay on Meadow Creek Road, then continues for 10.1 more miles to intersect Moyie River Road. This is a key intersection: Take the right, go just 0.3 mile, then turn left to the Meadow Creek Campground. To reach Placer Creek, go back to that junction, stay right, cross the bridge in 0.3 mile, and stay right onto CR 34-1/Deer Ridge Road. Follow this road for 2.1 miles to where it bears sharply to the right; there is a spur that leads to the left, with a gate before the bridge over Placer Creek. To reach the bridge site, go back to Moyie River Road but do not go south, back across the bridge. Instead, head north. You will see a faint dirt track that leads to the water and a decent camping spot. Finally, to reach Moyie Crossing, resume north on Moyie River Road for 7.7 miles and look for a left turn across the railroad tracks. If you are here during the right season, the gate will be open and you can drive in; we had to walk, but it was a short jaunt.

This picturesque primitive camping spot on the Moyie River has good bedrock and crumbling quartz veins nearby.

Prospecting

This area of the Panhandle region is full of mineralization, although much of it seems to be zinc, copper, and base metals. The Moyie River gives up good colors, however, and the area around Placer Creek is by far the best. If you look on maps, the source of the area's garnets is obvious—Ruby Mountain and Ruby Ridge dominate the skyline to the east. (Note that the infamous Ruby Ridge is actually southwest of Bonners Ferry.) At any rate, Placer Creek has good colors, and it empties into the Moyie River right across from Meadow Creek Campground. The upper Placer Creek site listed will also work if it is too crowded in the campground, and the road connects yet again with Placer Creek even higher up. I liked the bridge site listed here, as you can combine placering with a little lode sampling. There is a nice quartz seam exposed on the north side of the bedrock outcrop, and it has the characteristic rusty staining you like to see from decomposing sulfides. You can even try crushing up a sample and panning it. The Moyie River Crossing does not have great gold, but it contains some historical significance as a washed-out bridge. It also features an interesting outdoor sculpture by Jeffrey Funk titled *Kaniksu Passage*. He balanced large slabs of angular argillite on rounded granite and surrounded it with giant salmon circling patiently. It is a nice picnic area for the kids to blow off some steam. Copper Falls, at the top of Moyie River Road and then back up NF 2517, is well marked and worth a hike.

68 Wallace

See map on page 161.
Land type: Creek
County: Shoshone
GPS:
A - Prichard Creek: 47.65781, -115.97064
B - Eagle Creek: 47.66158, -115.87524
C - Burke: 47.52969, -115.79114
D - Sierra Silver Tour: 47.47173, -115.92559
E - Museum: 47.47165, -115.92472
F - Placer Creek: 47.43365, -115.90908
Best season: Spring–fall for outdoor sites
Land manager: Coeur d'Alene National Forest
Material: Fine gold, small flakes, black sands
Tools: Pan, bucket, hammer
Vehicle: Any; 4WD suggested for exploring
Special attraction: Sunshine Miners Memorial
Accommodations: Eagle City RV Park; seasonal campgrounds along Coeur d'Alene River; dispersed camping along the river as well
Finding the sites: Use Wallace as your headquarters for this region; it is right on I-90, which crosses the Idaho panhandle. To reach the mouth of Prichard Creek, leave Wallace by heading north on Sixth Street; it crosses under the interstate as 9 Mile Creek Road/NF 456. Drive 10.2 miles north, where the road turns into Beaver Creek Road, and go another 5.8 miles. You'll cross the river and reach Coeur d'Alene River Road; turn right and go 1.7 miles, then turn right onto Prichard Creek Road and park next to the water. To reach Eagle Creek, continue east on Prichard Creek Road for 2.7 miles, then turn left onto Eagle Creek Road/NF 152 and go 2.5 miles, or as far as you dare—this road washes out frequently. While you are in the neighborhood, drive on into Murray and check out a neat old mining town. Backtrack to Prichard Creek Road and continue east just 2 miles. To reach Burke from Murray, you can backtrack all the way to the Prichard Creek locale, then back to Wallace, or take the Kings Pass Road 2.8 miles until it becomes Delta Murray Road, then drive another 3.1 miles to join Beaver Creek Road and drive 12.7 miles back to Wallace. From Wallace drive east on Bank Street/Residence Street

One of the remaining buildings at Burke, Idaho. This small valley north of Wallace was a leading producer in the Silver Valley.

0.5 mile, go under the freeway, then up ID 4/Burke Road for 6.3 miles, where the road becomes NF 7623, and go another 1.5 miles to the power station. Stay left, cross the creek on Washington Basin Road, and circle around so you can reach Canyon Creek. The Sierra Silver Mine Tour and the Wallace District Mining Museum are within a block of each other in downtown Wallace at Fifth and Bank Streets. Finally, reach the Placer Creek locale by driving west on Bank Street to where it becomes King Street, then head south 0.4 mile, where the road becomes NF 456/ Placer Creek Road. Follow it for 3.3 miles to Hord's Ranch Road, where a bridge crosses the creek, and you can park safely there.

Prospecting

Wallace is the headquarters for the Silver Valley, where billions of dollars' worth of silver emerged from the surrounding hills. My records show GPS coordinates for over twenty-five locales up here, but these are the highlights and will give you an idea of the valley. This old mining district has been the scene of labor strife, boom times, reinvention, stock swindles, rampant speculation, mining tragedy, and much more. If you do not have time to explore the surrounding hills, the one spot I recommend is Site E, the Wallace District Mining Museum. Staff members are encyclopedic in their knowledge of the surrounding area, and the museum has an excellent selection of gifts and books and superb displays of local geology, rocks and gems from the mines, mining equipment, and much, much more. Then, if you have the time, the Sierra Silver Tour (Site D) is excellent family fun. If you like tours, there is a second one: the Crystal Gold Mine, at 47.4337, -115.9091. You could spend quite a bit of time up here and have a great visit. The panning at the mouth of Prichard Creek is excellent; once you park safely, follow the creek to the North Fork Coeur d'Alene River and sample the bar forming downstream from the creek. At Site B, known locally as Oregon Gulch, the road is washed out, and to proceed up Eagle Creek you'd need to make the ford. Don't bother though; it's claimed all along that stretch. Along the way to this site, Eagle City Park is worth checking out at 47.6468, -115.9111, offering fee-based panning. The Burke panning spot is more challenging for getting color, but there are extensive tailings piles and mine dumps you can bring a hammer to. If nothing else, it's a good excuse to drive up to Burke and look it over. Finally, Placer Creek at Site F is reliable for color, and the only claims listed for this section of the creek are for lode mines, so it is open to explore here.

Other attractions include the Sunshine Miners Memorial, located at 47.5281, -116.0479. Ninety-one miners tragically died in an underground fire at the mine in 1972. You can see the actual Sunshine Mine, still in operation, at 47.5018, -116.0714.

69 St. Joe River

See map on page 161.

Land type: Creek
County: Shoshone
GPS:
A - Bluff Creek: 47.19101, -115.49791
B - Conrad Crossing CG: 47.15815, -115.41599
C - Gold Creek: 47.1509, -115.40742
D - Spruce Tree CG: 47.037731, -115.34897
E - Heller Creek CG: 47.06415, -115.21924
Best season: Late summer
Land manager: St. Joe National Forest
Material: Fine gold, small flakes, black sands
Tools: Pan, shovel, bucket
Vehicle: Any for sites A–D; 4WD strongly suggested for site E
Special attraction: Red Ives Ranger Station
Accommodations: Multiple developed campgrounds, including Gold Creek, Tin Cup, Conrad Crossing, and Heller Creek; dispersed camping along St. Joe River
Finding the sites: The small town of Avery sits on the St. Joe River, about 46 miles from St. Maries. From Avery take St. Joe River Road/NF 50 about 21.3 miles to reach a slim pull-out on the inside bend of the river—this is the Bluff Creek locale. Conrad Crossing Campground is 6.9 miles farther up the road. The junction with Gold Creek Road/NF 388 is just 0.8 mile farther. To reach Spruce Tree Campground, continue on St. Joe River Road for 9.6 miles to Red Ives Ranger Station, then turn onto NF 218/Spruce Tree Campground Loop for 1.7 miles. To reach Heller Creek Campground, turn left onto NF 320 at Red Ives and drive a very tough 11.8 miles.

Prospecting

All rockhounds and prospectors have that one place that they just cannot seem to adequately investigate—Heller Creek was my "Moby Dick." I tried to come in from Avery one year and faced a washout. Another year I tried to come in from the Montana side, leaving I-90 at Superior and journeying up via Cedar Creek Road, only to run into deep snow at Missoula Lake. But in 2022 I broke through and drove right to it from Avery. There is good gold up

A good lesson in why you should call ahead when planning trips to the Idaho backcountry: A washout blocked entry to the upper St. Joe River.

here, and I was panning color on the way up and over on the Montana side, too. The St. Joe Placer Mine at 47.0503, -115.1967 isn't easy to get to. The Spruce Tree Campground has a hiking trail to get closer to the old Ruby and Joe Placer claims about 2.5 miles up the St. Joe River from camp. Meanwhile, I did pan some color at the mouth of Gold Creek but not up the creek itself. At Conrad Crossing Campground, there are a couple prospects in the vicinity, although the Whitetail Mine was mostly a copper prospect, said to be directly across the river from the campgrounds. I had colors in my pan at Bluff Creek as well, although I wished for lower water to find better hand specimens related to a couple of prospects rumored to be in that area. Because the St. Joe River is protected as Wild and Scenic, you won't be able to use anything but a pan and screens, you can't dig into the bank, and you must be extra careful to fill in all your holes.

70 Hoodoo Placers

See map on page 161.
Land type: Creek, riverbank
County: Latah
GPS:
A - Poorman: 46.95721, -116.61166
B - Cleveland Gulch: 46.98615, -116.55617
C - Mary Lee Placer: 47.00752, -116.53822
Best season: Late summer
Land manager: St. Joe National Forest
Material: Fine gold, small flakes, black sands; garnets
Tools: Pan, sluice, highbanker, dredge
Vehicle: Any
Special attractions: Emerald Creek Garnet Area, Fossil Bowl fee-dig site
Accommodations: Emerald Creek and Laird Park CGs; dispersed camping along the Palouse River
Finding the sites: Use the small crossroads of Harvard, Idaho, as a base—it is at the junction of ID 6 and ID 9, roughly 30 miles northeast of Moscow. From Harvard drive northeast on ID 6 about 3.6 miles to Palouse River Road/NF 447 and turn right. Go about 3.7 miles east and look for the road heading to the river at the junction. This is as close as possible to the Poorman location. To reach the next two spots, return to Palouse River Road and head east 2.4 miles, then turn left to head up NF 767, which follows the North Fork. Drive 1.5 miles, then look for good parking to reach the water. This is approximately the location of the old Cleveland Gulch area. Drive another 2 miles up the road to reach the Mary Lee Placer.

Prospecting

This area is rich in mining history, particularly during the Depression years of the 1930s. The Hoodoo District was primarily a placer area, and you will see plenty of tailings. Be on the lookout for claim markers and note that the local timber company is continually experimenting with offering recreation and camping here on a lease basis. My readings should keep you clear of the claimed areas, and there are fewer claims up here than in the past. At the Poorman site there is easy access to the creek; you might find other access

Someone put a lot of work into this structure, either to winter over or to protect gold or explosives. It is located where the North Fork of the Palouse River joins the main stem.

points before this. You'll probably see lots of campers throughout this area during peak summer months, as there is so much room to spread out, and it's flat enough to accommodate trailers. Site B at Cleveland Gulch is mostly tailings piles, with some water access, and the same holds for the Mary Lee Placer locale. I have seen dredges up here that did well, but it is a long way to bedrock in most places, and you are sure to be working through old tailings unless you can spot areas where the tailings pinch out. Be on the lookout for garnets too; this area is well known by rockhounds. The nearby Fossil Bowl fee-dig area is popular with families—for a modest fee, you will surely go home with at least partial leaf fossils, and sometimes more. The famed Emerald Creek Garnet Area is another fee-dig operation where you can recover prized star garnets; check if the Forest Service has reopened it after COVID.

71 Pierce

See map on page 161.
Land type: Creek
County: Clearwater
GPS:
A - Orofino Creek: 46.50558, -115.88254
B - Armstrong Gulch/Orofino Creek: 46.45469, -115.76021
C - Orogrande Creek: 46.58284, -115.56276
Best season: Late summer
Land manager: Clearwater National Forest, BLM–Coeur d'Alene
Material: Fine gold, small flakes, black sands; red garnet, green epidote, clear sapphire (Orofino Creek)
Tools: Pan, sluice, highbanker, dredge
Vehicle: Any; 4WD suggested—roads are decent but rough in stretches.
Special attractions: Pierce discovery site and museum
Accommodations: Hollywood CG near Pierce, Weitas Creek CG on North Fork of the Clearwater; dispersed camping along French Creek, Orogrande Creek, and elsewhere on USFS land
Finding the sites: Pierce is a good headquarters for this region, as it has gas and groceries. To reach Orofino Creek, drive north on ID 11 from the gas station at the north end of town about 3.9 miles and turn left, across from the ruins of the old mill, when you reach the small ponds. This road curls back against ID 11 and down to Orofino Creek after 5.1 miles. To reach Armstrong Gulch, return to Pierce, but this time head south on ID 11 for 0.8 mile and look for a left turn onto French Mountain Road, which follows Orofino Creek. After about 3.2 miles you will notice a good spot to access the creek. To reach Orogrande Creek, follow French Mountain Road and go 20.2 miles farther up.

Prospecting

Pierce was the site of the first big gold discovery in what became the state of Idaho, in 1860 at Canal Gulch by E. D. Pierce and Wilbur F. Bassett. There is a nice kiosk up by the gas station, and the J. Howard Bradbury Logging Museum has a little mining history tossed in. The Cedar Inn Restaurant may still carry mining supplies; if so, they are a wealth of local information. Much of the early mining was right downtown in Pierce and along Canal Gulch, but

Pierce, Idaho, is justifiably proud of its history as the first major gold strike in what was then still part of the Washington Territory, in 1860.

that is all private now. Poormans Creek and Cow Creek, northwest of town and downstream, also had considerable activity, if you are exploring. The site on Orofino Creek is just 0.5 mile downstream from big placer diggings on Orofino Creek, and if you can angle farther upstream on the logging roads without encountering timber company postings, you could do well. There are reports of garnet, epidote, and sapphire here in Orofino Creek, but you will probably need a hand lens to look at your concentrates to find such exotics. The Rosebud Creek site is just off Orofino Creek and should get you off any patented land. One year we saw postings and one year they were gone, so keep an eye out. The stretch of Orofino Creek from Pierce to the US Forest Service boundary saw considerable activity in the early days of the district, as did Armstrong Gulch and other areas once you clear the boundary. We did well up to a couple miles above the Rosebud site. This stretch of road leading up to the Orogrande Creek site is not horrible, and it is shorter than going back to Pierce.

The French Creek Placer operation was at 46.5310, -115.6502, and there have been GPAA claims at 46.5659, -115.6193. If you come in from Pierce, there were placers on Crystal Creek, Irish Creek, and Elk Creek, among others. Note that you can continue on this road to reach the North Fork of the Clearwater River, and there are more placer areas between you and the mouth of Orogrande Creek, which is also a great spot at 46.6310, -115.5077, about 6 miles farther north. You can reach Superior, Montana, via this route, over Hoodoo Pass; it is about 102 miles from Pierce to Superior. There are dozens more placer mines reported up on the North Fork.

72 Lolo Creek

See map on page 161.
Land type: Creek
County: Clearwater
GPS:
A - Bridge: 46.31296, -115.75021
B - GPAA claims: 46.30954, -115.75374
C - Campground: 46.29314, -115.75242
D - Lower: 46.28848, -115.75889
Best season: Late summer
Land manager: Clearwater National Forest
Material: Fine gold, small flakes, black sands
Tools: Pan, sluice, highbanker, dredge
Vehicle: Any
Special attraction: Lolo Pass
Accommodations: Lolo Creek has multiple campgrounds; dispersed camping all along creek as well
Finding the sites: To reach Lolo Creek, drive south from Pierce on ID 11 for 0.9 mile and take a left onto Fromelt Road. Go 0.4 mile to turn right onto Browns Creek Road, then after 7.5 miles turn left onto Musselshell Road. After 5.5 miles you will reach Lolo Creek Road. Turn right (south) and drive 3 miles to the bridge coordinates. The GPAA claims start about 0.4 mile below here and continue for 0.5 mile. The campground is about 1.8 miles south of the bridge coordinates, and the final lower spot is another 0.5 mile south of the campground. If you continue south, there are more interesting areas worth checking.

Prospecting

This area yields decent colors all the way up to Hemlock Butte, but that is a tough drive, and you are better off attempting to access the creek from French Mountain Road and CR 535. This stretch of Lolo Creek is paved and easy to find, with excellent developed and dispersed camping, easy access to the water, and plenty of wildlife. Moose are abundant out here, and there are even reports of Bigfoot sightings, if you are a believer. There are several reported lode deposits up by Hemlock Butte, but not along the lower area, and none

Keep an eye out for wildlife as you drive the back roads at Lolo Creek. We have seen moose every time we visited the area.

were big producers. Still, the creek offers good prospecting, with plenty of inside bends containing lots of gravel if you come when the water is low. The GPAA claims include much of Brady Creek, which surprised me, until I spotted what looked like old tailings piles along its banks. It makes you wonder if more of the creeks in this area contain good color, especially as you go up. If you spot new claim markers, consider sticking to the campground and picnic area.

CENTRAL IDAHO

Central Idaho West

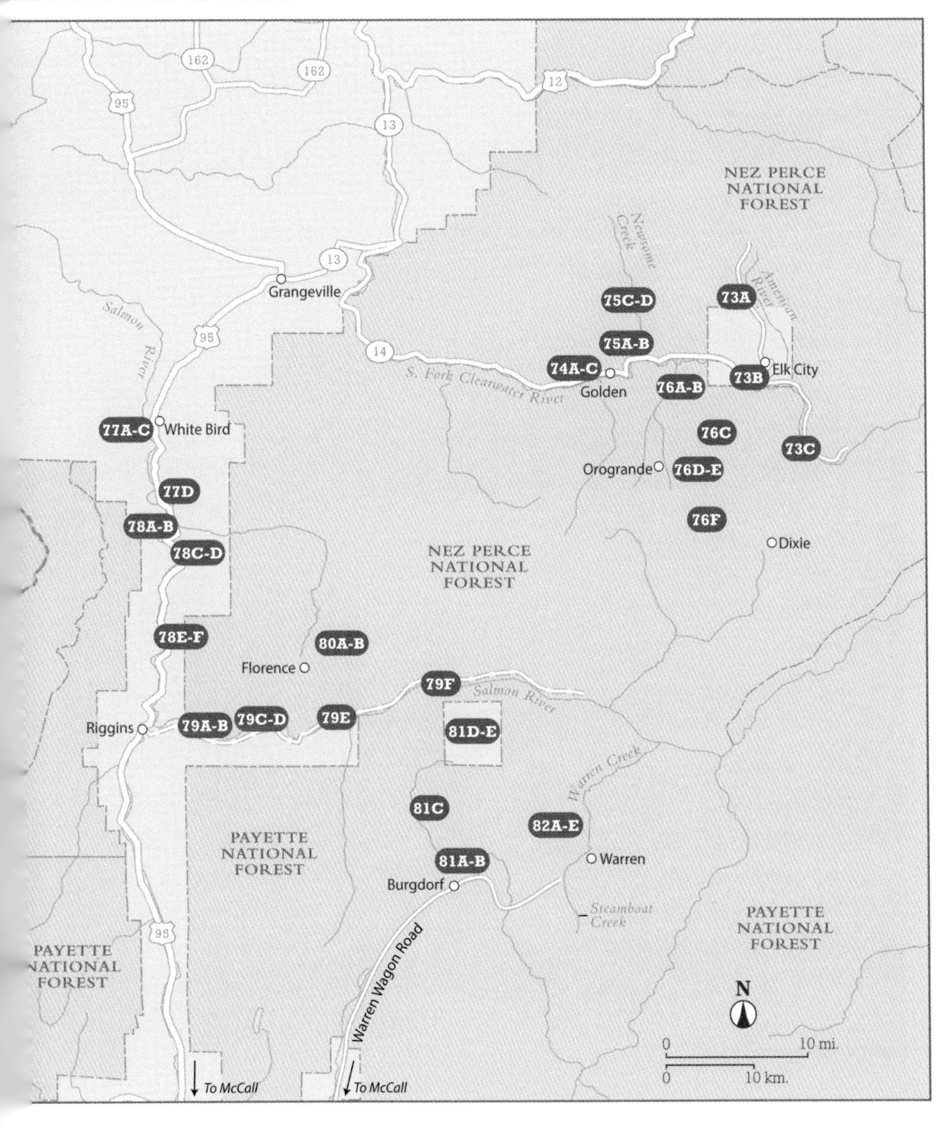

73 Elk City

Land type: Creek, river
County: King
GPS:
A - American Placer: 45.88442, -115.44729
B - Bridge: 45.80846, -115.47462
C - Gold Point: 45.72362, -115.37848
Best season: Late summer
Land manager: Nez Perce National Forest
Material: Fine gold, small flakes, black sands
Tools: Pan, sluice, highbanker, dredge
Vehicle: Any; 4WD suggested
Special attraction: Red River Hot Springs

If you have time, check out rustic Red River Hot Springs, east of Elk City. According to reports, new ownership has restored their quality.

Accommodations: Developed campground at Newsome Creek and Crooked River; dispersed camping on USFS lands above American site

Finding the sites: Elk City is about 51 miles east of Grangeville via ID 14. To reach the American Placer locale at Site A, take the Elk City Wagon Road off ID 14 a short distance to Elk Creek Road and turn right. Go 3.2 miles north, and then turn left onto Erickson Ridge Road/Falls Point Road. Go 0.9 mile, then continue on NF 443 about 1.1 more miles. This is the edge of USFS land. To reach the bridge site, backtrack on ID 14 from Elk City about 2.8 miles and locate the bridge where the Red River flows into the Clearwater. The Gold Point site is about 12 miles up Red River Road, about halfway to the rustic hot springs.

Prospecting

The Elk City area is full of good gold, but you will have to dodge private land and claims to work up there. The Gold Point site is set aside for panning. If you can find open spots along the Red River, test them out, as this was a good producer back in the day, with multiple operations. You should see several access spots, and if there are no signs, you might give any inside bend or gravel bar a test. The American site is in the heart of good placer ground, as the hills northeast of Elk City are dotted with old placer diggings and prospects. Look for good access prior to reaching the coordinates, as there are some easy access points that come and go as claimed spots. The bridge over the Red River shows good color as well, but you will want to stick close to the easement. This was the site of the Cal-Idaho Placer operation.

74 South Fork Clearwater River

See map on page 175.
Land type: Creek
County: King
GPS:
A - Wide spot: 45.79757, -115.71935
B - Boulders: 45.80796, -115.80133
C - NF 469: 45.81687, -115.81979
Best season: Late summer
Land manager: Nez Perce National Forest
Material: Fine gold, small flakes, black sands
Tools: Pan, sluice, highbanker, dredge
Vehicle: Any
Special attraction: Elk City
Accommodations: Developed campgrounds along South Fork of the Clearwater, such as Castle Creek, South Fork, and Leggett Creek; dispersed camping throughout this area
Finding the sites: Golden is on ID 14, about 33 miles east of Grangeville via Mount Idaho Grade. From Golden, drive east for 2.3 miles to reach Site A. Site B is 7.1 miles east of Golden. Site C is 8.2 miles east.

Prospecting

The South Fork of the Clearwater has multiple access spots, so you just have to dodge current claims to find a nice place to prospect. Because you are below four major districts—Elk City, Red River, Orogrande, and Newsome Creek—you should easily produce color. You will probably dodge anglers here, as this is a famed trout area. However, with ample camping, good color, and easy access, this area is a welcome change from the "billy goat" trails and dusty roads that many Idaho districts contain. The South Fork is scenic and uncivilized, but officials have not designated it as a Wild and Scenic River, so dredging is allowed, and that means plenty of placer claims exist, especially between Golden and Elk City. The stretch between Golden and Castle Creek Campground is still open; more good spots can be found north of Cotter Bar,

This stretch of the South Fork of the Clearwater River offers good access to clean, gold-bearing gravels.

and there are no active claims there, either. Consult the Idaho Department of Lands at https://www.idl.idaho.gov/mining-minerals/dredge-placer-mining/ for updated information on dredging seasons. For panning, you should be able to move rocks and start a good hole. Just be sure to fill it back in.

75 Newsome Creek

See map on page 175.
Land type: Creek, cliffs
County: Idaho
GPS:
A - Newsome Bridge Road: 45.82977, -115.60632
B - Moose Creek hydraulic mine: 45.83398, -115.60159
C - Nugget Creek: 45.87996, -115.61754
D - Sing Lee CG: 45.88441, -115.62609
Best season: Late summer
Land manager: Nez Perce National Forest
Material: Fine gold, small flakes, black sands
Tools: Pan, sluice, highbanker, dredge
Vehicle: Any; 4WD suggested
Special attraction: Asbestos Falls
Accommodations: Multiple campgrounds; Sing Lee is a good central location. Dispersed camping throughout this area on USFS land.
Finding the sites: From Golden drive 5.7 miles east and look for Newsome Bridge Road. At the bottom of the road, there is a giant parking area with good access to the Clearwater. To reach the Moose Creek hydraulic mine, go up Newsome Bridge Road 0.3 mile and look for a sharp right onto NF 307. Drive about 0.7 mile and look for a gated road heading up to the left. Walk in about 0.1 mile. To reach the mouth of Nugget Creek, return to Newsome Bridge Road and drive 2.8 miles up (north), then continue onto the Old Elk City Wagon Road and go 1.4 miles. You will see access to Newsome Creek on the left and a logging road to the right. Sing Lee Campground is another 0.5 mile up the main road.

Prospecting

Newsome Creek saw considerable placer mining in its heyday, with hydraulic mines dotting the cliffs around here. The Moose Creek hydraulic mine at Site B is a splendid example of what the teams did, and you can still pan good color out of the creek that drains it. The actual mouth of Newsome Creek hosts a claim, and we saw a dredge one time we visited. While in this area, explore up Leggett Creek as well—there were more mines up there, including

Sing Lee Campground at Newsome Creek offers a few amenities and access to gold-bearing gravels on Newsome Creek.

a giant hydraulic mine. The two big quarries up Leggett Creek Road/NF 649 show you are on the right trail. Nugget Creek at Site B is a reliable producer as far up as you care to hike it—don't take the trail, but follow the creek instead. We found an old Chinese "apartment" complex near the logging road, just above the creek, and we hiked this road at least 0.5 mile up and still found good color throughout. The Sing Lee Campground is an easy base of operations, and there is reliable color in Newsome Creek here. If possible, camp along the water rather than across the meadow near the outhouse. There were more placer mines up NF 1858, but like most of this area, you'll have to dodge claims that pop up when the price of gold is high.

76 Orogrande

See map on page 175.
Land type: River and creek bank, tailings
County: Idaho
GPS:
A - Tailings: 45.79632, -115.53192
B - Trout Group claims: 45.79146, -115.55379
C - Relief Creek: 45.74727, -115.51401
D - Fivemile CG: 45.71691, -115.54041
E - Butte & Orogrande Mine: 45.70815, -115.54055
F - Homestake: 45.66344, -115.52895
Best season: Late summer
Land manager: Nez Perce National Forest
Material: Fine gold, small flakes, black sands
Tools: Pan, sluice, shovel, bucket
Vehicle: Any; 4WD suggested
Special attractions: Gnome Mill and buildings
Accommodations: Multiple developed campgrounds and dispersed camping along the Crooked River
Finding the sites: From Golden on ID 14, drive 11.7 miles east to the mouth of the Crooked River and cross the bridge there onto Crooked River Road/NF 233. Drive 2.2 miles to find a camping area with access to the Crooked River. This is Site A. There is a nice bend on the river near the old Trout Group claims about 1.7 miles farther up the main road (Site B). Drive another 4.2 miles up the road to a left turn and go just 0.2 mile to a nice pull-out on Relief Creek at Site C. About 2.5 miles farther up, you will reach Fivemile Campground, with access to water. Another 0.6 mile up, you will see the remains of the Butte & Orogrande Mine, which is Site E. Finally, about 1 mile farther, you have a choice of heading for the Buffalo Hump district (which unfortunately is mostly private and reachable only by trail), backtracking to ID 14, or continuing around the loop to Dixie and then back to Elk City. We took the loop and enjoyed it immensely, gathering good ore samples at the Homestake site. From this junction bear right on Orogrande-Dixie Road and drive about 3.6 miles. You will be in some tight hairpin turns; the fourth one has some scratchings that indicate the mine site.

Do not count on services at Orogrande; it is off the grid. Try the loop all the way around to Dixie and back to Elk City if you want a true Idaho adventure.

Prospecting

The tailings along the Crooked River attest to the amount of effort early miners expended here, and they did quite well. The easy gold is gone, but you should be able to produce colors if you can get to a good spot on the river. The tailings piles themselves are worth checking with a metal detector. We did well at the Relief Creek spot at Site C, and there are more prospects up that drainage. Some interesting old buildings were moved from the Gnome Mill to a site along the road, and they are worth photographing. Orogrande itself is fairly busy during the summer, and visitors are discouraged from damaging the structures. We found some good hand samples at the Butte site, but do not be surprised to see claim signs, as that seems to be the trend. Fortunately for gold panners, most of the claims here are lode claims, which means you can still pan and sample the tailings. There were no claim signs up when we visited, but that could have been due to someone maliciously tearing them down. Access to the Buffalo Hump district is difficult, if not outright restricted, which is a pity. It showed early promise. One site to look for just south of Orogrande is the Petzite Mine at approximately 45.6875, -115.5311. Petzite is a steel-gray to sooty-black mineral that looks better when it tarnishes bronze. It is a rare telluride—the chemical formula is Ag_3AuTe_2, so it is rich in silver and gold. That one is on my to-do list when I get back there. When we were at the Homestake prospect, someone had done some trenching and we found good samples in the piles along the trenches. It is about 23 miles from Orogrande to Dixie, and then another 30 miles to Elk City. While we did enjoy a cold beverage at Dixie, we were not successful in gathering much good local information. All the land around Dixie seems to be private or claimed, and we needed more time to explore the mines farther from town.

77 White Bird

See map on page 175.

Land type: Riverbank
County: Idaho
GPS:
A - Lyons Camp Rd.: 45.76009, -116.32474
B - Hammer Creek: 45.76193, -116.32532
C - Launch: 45.73789, -116.31542
D - Horseshoe Bend: 45.66479, -116.28973
Best season: Late summer
Land manager: Idaho Dept. of Fish and Wildlife, BLM–Cottonwood
Material: Fine gold, small flakes, black sands
Tools: Pan, sluice; dredging by permit
Vehicle: Any; 4WD required for Lyons Camp Road
Special attraction: White Bird National Battlefield
Accommodations: Numerous developed campgrounds along the Salmon River; dispersed camping at some of the larger bars
Finding the sites: White Bird is east of US 95, about 17 miles south of Grangeville. Look for a large bridge crossing White Bird Creek, and just south of the bridge, turn east onto Everest Street. Go into town, turn left onto Main Street, and head southwest. Go under the bridge and about 0.8 mile to a right turn onto Lyons Camp Road. This one is nasty. Go about 0.8 mile to a small road leading to the water and a nice beach area. Work among the boulders there. To reach Hammer Creek Campground, return to Lyons Camp and go south about 0.9 mile to the bridge, cross the Salmon River on Doumecq Road, and swing right (north) about 1.5 miles. Watch for a nice U-Pick orchard along the way with excellent apricots, apples, and plums. You do not need to go all the way into the campground; the best spot is upstream of the camp near some bedrock, almost directly opposite the spot on Lyons Camp Road. To reach the boat launch, return to the bridge and go right on Old Highway 95. The launch is about 0.6 mile. To reach the old boat launch on Horseshoe Bend, go 0.3 mile to US 95, turn right (south), and drive about 6 miles. You will pass a nice spot at Skookumchuck, and you will see big diggings at Taylor Bar. There does not seem to be any good access at Roby Bar, but Horseshoe Bend has a couple spots. There is a fishing area along the north bridge, and another one on the south bridge. Try those, or turn left onto Rivers Bend Road and hug the water for 0.2 mile.

Columnar basalt bedrock on the Salmon River, just above Hammer Creek Recreation Site. This photo is from Lyons Camp Road, on the east side of the river.

Prospecting

This area is famous for being part of the journey of the Nez Perce when they were fleeing US troops from Oregon to Canada, and they fought a lengthy running battle at White Bird. The Salmon River has excellent gold, and this stretch of the river is open to the public. I haven't dredged here, but the panning is very good. The Lyons site is remote, and the boulders I worked contained excellent black sand and good color in every pan. The Hammer Creek spot is also good, and if you can make your way to the bedrock, you should do well crevicing. The BLM boat launch is average for color, especially on the upstream end of the bar. At Horseshoe Bend the river has tried to carve a channel when the water is running high, and there is good color here, so you don't have to wonder if the actual Horseshoe Bar Placer just downriver is a lot better. There are undoubtedly more good bars and bends on the river below here, and you could drive all the way to Pine Bar at 45.8902, -116.3314 to check a truly remote site. Continue on to the Blackhawk Bar sites for more Salmon River access south of here.

78 Blackhawk Bar

See map on page 175.
Land type: Riverbank
County: Idaho
GPS:
A - Rest area: 45.64495, -116.29211
B - Blackhawk Bar: 45.62755, -116.30218
C - Long Gulch: 45.61256, -116.27843
D - Maynard Hole: 45.59715, -116.27324
E - Spring Bar: 45.51778, -116.30502
F - Carver: 45.51178, -116.29805
Best season: Late summer
Land manager: Idaho Dept. of Fish and Wildlife, BLM–Cottonwood
Material: Fine gold, small flakes, black sands
Tools: Pan, sluice; dredging by permit
Vehicle: Any; just be careful if leaving the road to get closer to water.
Special attraction: Riggins
Accommodations: Numerous developed campgrounds along the Salmon River; dispersed camping at some of the larger bars
Finding the sites: From White Bird on US 95, drive south 9.4 miles. There is a large rest area with a prominent bedrock outcrop on the southern end. Blackhawk Bar is about 2.1 miles farther south. Look for the middle of the bar, with a nice set of rapids with large boulders to work around. The old meander at Long Gulch is about 1.7 miles farther south, with bedrock on both sides of the road. Maynard Hole is 1.2 miles farther south; there is a boat launch and nice beach areas. Spring Bar is another 6.4 miles south, past more good access spots, then past Lucile, to a public boat launch. There is a rough road down to the upper bar, or you can walk down. Finally, another 0.5 mile south you will reach a wide pull-out area. This is Carver. The dirt road leading back north along the water is tempting, but the river is very straight here, and the bedrock outcrop is a far better spot to sample. This spot is 7.5 miles north of Riggins.

This stretch of the Salmon River near Blackhawk Bar offers multiple access opportunities to the river.

Prospecting

This is classic river prospecting: You can spend all day digging a big hole, or you can sample your way along at your leisure. My favorite spot here is Blackhawk Bar, where I found large flakes in the first pan that I dug from around the big boulders at the water's edge. When the water is low, these areas are all worthwhile, but if you come in the wrong month, you'll be lucky just to see flour gold, even if you try around roots or moss. Your best bet is at low water, around bedrock or large boulders, and at the top of inside bends. The river is a rushing torrent during flood stage, which you should avoid, but during the months when the water is low, anywhere you can find traps, buildups, or crevices, you will do well. Be sure to fill in your holes.

79 Ruby Rapids

See map on page 175.
Land type: Riverbank
County: Idaho
GPS:
A - Island Bar: 45.41599, -116.25835
B - Second Bar: 45.41109, -116.24827
C - Ruby Rapids: 45.40511, -116.19447
D - Spring Bar CG: 45.42306, -116.15745
E - French Creek: 45.42497, -116.03043
F - Chittam Bar boat ramp: 45.45927, -115.89243
Best season: Late summer
Land manager: Nez Perce National Forest
Material: Fine gold, small flakes, black sands; garnets at and below Ruby Rapids
Tools: Pan, sluice, shovel, bucket
Vehicle: Any if staying on road and driving carefully; 4WD suggested for closer water access
Special attraction: Riggins Hot Springs
Accommodations: Multiple campgrounds along this stretch of the Salmon River; dispersed camping at many access points
Finding the sites: From the intersection of Lodge Street and Main Street in Riggins, drive south on US 95 for 0.8 mile to the Big Salmon Bridge. Turn east and cross the Salmon in about 0.1 mile, then continue 3.8 more miles to the ruts that lead to the water. The next bar access is another 0.7 mile up the river. Ruby Rapids is 3.1 miles farther up; you will cross the river about halfway there. Continue another 2.5 miles to reach the beginnings of Spring Bar; the campground is within view. French Creek is another 8.2 miles farther up, and the end of the road at Chittam Bar is 7.8 miles after that.

Prospecting

At Site A and Site B, the two bars close to Riggins, you'll have great access to the top of the gravels, where heavies sink out first. At Ruby Rapids you will find numerous garnets, some as big as peas. The bedrock above you is full of garnets, so if you want more of the tiny crystals, move below that rock. The

One of many extensive gravel bars on the Salmon River near Ruby Rapids. You will not find rubies—the red crystals in the sands here are garnets.

problem at Ruby Rapids is parking—you would be better off having someone drop you off than trying to wedge your vehicle between the road and the cliff. We enjoyed the campground at Spring Bar, which is Site D, but the panning was also good a little farther down at a popular access point. Good sandy beaches do not always mean good gold; white sands are typically lighter and move faster with the river than the concentrates that trap in bedrock and under boulders, so be prepared to dig deeper. At Site E at French Creek, you will find those boulders and bedrock, and the creek mouth has excellent colors in the pan. There is a nice footbridge across the Salmon River at the boat launch at Wind River; note that if you go up NF 318 here, you can reach the Marshall Mountain district and Burgdorf Hot Springs, even Warren or McCall, depending on your needs. Site F at Chittam Rapids, at the end of the road, has a good developing bar and excellent bedrock just upstream from the parking area and boat ramp.

80 Florence

See map on page 175.
Land type: Creek
County: King
GPS:
A - New Florence site: 45.50166, -116.03046
B - Hydraulic pits: 45.52431, -116.01331
Best season: Late summer
Land manager: Nez Perce National Forest
Material: Fine gold, small flakes, black sands
Tools: Pan, sluice, shovel, bucket
Vehicle: 4WD required
Special attraction: Riggins
Accommodations: Primitive camping throughout the area; developed camping near Riggins and along Salmon River
Finding the sites: From Riggins head east along the Salmon River on Big Salmon Road. This is NF 1614. The bridge is about 6.4 miles from US 95. Cross the bridge and pass the Ruby Rapids site; about 3.1 miles past the bridge, locate the road up Allison Creek. This is NF 221. After about 12.8 miles, turn left onto NF 394 and go about 5.7 miles. Take the turn onto NF 643, which is Florence Road. Good signs will guide you the final 6 miles to the townsite. Note that you can reach Florence via Grangeville, using Mount Idaho Road and Grangeville-Salmon Road to Florence Road. Finally, you can exit US 95 at Slate Creek from the north; it's a pleasant drive and fairly straightforward. Also, road numbers on the ground are at least thirty years old and may not match up with new, printed maps and Google Earth, but the signs point you in the right direction. I found this all very confusing, and I thought you should know going in that the signs work. To reach the hydraulic pits, continue north on NF 643 for 1.9 miles.

Prospecting

Fabulous Florence was a rich strike back in the 1860s, and miners flocked to the area to stake good ground. The earliest prospectors came from Pierce and tracked down unexpectedly rich ground in August 1861. Woefully unprepared for winter, the camp experienced a severe famine, but by spring, the

This New Florence sign gives you a good layout of the area. Try to find Black Sand Creek.

stampede was on. The rich claims soon were worked out, and the camp had only 575 men by 1863. Today scarcely a trace of the town remains among scars of the old diggings. There are still several active small operations, multiple active lode and placer claims, and a lot of private land to dodge, so be on the lookout for signs if you go exploring. Ozark Creek, Healy Creek, Baboon Gulch, Gross Creek, Meadow Creek, Imperial Creek, White Sand Creek, and Black Sand Creek all had placer operations at one time, and most of the roads in this area take you to prospects. The trick to panning up here is to scout out good access, which will take some time. You can always poke around tailings where miners dredged or hydraulicked, fortunately. The Florence Cemetery at 45.5103, -116.0295 has a new information kiosk, outhouse, and hiking trail, and does a good job of bringing this Wild West locale to life. The hydraulic pit at Site B shows the scale of the work done out here. There are extensive dredge ponds at 45.5155, -116.0065. The first time we visited, we came too early in the season, and winter blowdown made stretches of the road between Florence and the Salmon River impassable without a chain saw.

81 Burgdorf

See map on page 175.
Land type: Creek, mine tailings
County: Idaho
GPS:
A - Ruby Meadows: 45.23667, -115.87794
B - Secesh River: 45.26496, -115.85828
C - Lake Creek: 45.32713, -115.94297
D - Silver Summit Mine: 45.40568, -115.86906
E - Kimberly: 45.40418, -115.86749
Best season: Late summer
Land manager: Payette National Forest
Material: Fine gold, small flakes, black sands; corundum, garnets
Tools: Pan, sluice, highbanker, dredge, hammer
Vehicle: Any to Burgdorf; 4WD suggested, required for Marshall Mountain
Special attraction: Burgdorf Hot Springs
Accommodations: Developed campgrounds near Burgdorf; cabins for rent at the hot springs; dispersed camping throughout this region
Finding the sites: From a mile west of McCall, drive north on Warren Wagon Road for 28.1 miles. Look for the sign to Ruby Meadows, and drive in as far as you can. You'll find mostly snowmobile and ATV tracks, so you may not get far by vehicle. The Laughing Water Placer area is 1.1 miles in, and the ponds that mark Ruby Meadows are another 1.2 miles. The Secesh River site is easier—it's just 2.1 miles beyond the turn for Ruby Meadows, then look for dirt tracks to the right. To reach the Lake Creek Placer workings, return south to the turn for Burgdorf Hot Springs, take Burgdorf Road/NF 246, and go past Burgdorf, for a total of 5.2 miles from the turn off Warren Wagon Road. There is a turnoff to the right. Another 1.2 miles farther up is the turn for Marshall Mountain at Corduroy Meadows Road/ NF 318. Turn right and follow NF 318 for 9.4 miles. At the elbow turn heading left, you will go downhill and right. After 0.3 mile look for a sharp left turn ahead and take it about 0.1 mile to the Silver Summit Mine. Stay out of the buildings and concentrate on the tailings piles. Back at that sharp left turn, you can go straight just 0.1 mile to reach the Kimberly area. It is quite photogenic, but again, stay out of the buildings.

The tailings piles at the Silver Summit Mine are much more interesting than anything inside the building, so stay out!

Prospecting

Burgdorf Hot Springs is a nice base camp for trips throughout this region. The Ruby Meadows spot used to be open to drive all the way in, but you will need ATVs now, or foot power. It is worth the hike, though, with good gold and rare rubies and sapphires in the concentrates. The placer mine at the coordinates is an active claim, and at 2 miles one-way, it's a long ways for just a pan or two. About halfway there you'll see considerable scars that are *not* claimed, at about 45.2483, -115.8919, reached via a slight left from the main trail. This is the Laughing Water Placer. The Secesh River locale at Site B is not under claim; ideally, you could access the mouth of Ruby Creek at 45.2578, -115.8788. This is a straight stretch of the river, draining Ruby Meadows, but there are some big boulders. The Lake Creek site was a large placer operation at one time, and there are plenty of scars and tailings to help you find it. Willow Creek has little water, but you can make your way to Lake Creek and work there. The Marshall Mountain district has seen sporadic activity lately, and many old buildings remain. The road in is rough, and the roads in the district are even worse. There are excellent tailings to explore at the Silver Summit Mine, so bring a hammer. You can try panning in the creeks here, but they are very small. The Kimberly Mine is on the far end of Kimberly Lake, with a small footpath around the lake reaching it. If you drive farther through the Kimberly camp, there are additional prospects. The Golden Anchor, Sherman-Howe, and Old Kentuck locales are also worth exploring.

82 Warren

See map on page 175.
Land type: Creek, mountains
County: King
GPS:
A - Steamboat Creek: 45.26144, -115.71661
B - Bridge: 45.27521, -115.69687
C - Crossing: 45.28048, -115.70206
D - Dredge remains: 45.29086, -115.69608
E - Charity Gulch: 45.248741, -115.654767
Best season: Mid- to late summer
Land manager: Payette National Forest
Material: Fine gold, little black sands
Tools: Pan, sluice, highbanker, dredge
Vehicle: 4WD required
Special attraction: Warren
Accommodations: Developed campgrounds near Burgdorf; dispersed camping throughout the area, on Steamboat Creek and Warren Creek outside of town
Finding the sites: From just west of McCall, turn north onto Warren Wagon Road past the lake, past the turn for Burgdorf, through Secesh, and onto Steamboat Creek, a total of 42.2 miles. You will see a bridge across Steamboat Creek, which is also a good place to stop and check, or continue on to the coordinates, the site of the Dreadnought Placer. Continue on Warren Wagon Road for 1.5 miles to reach the bridge over Warren Creek. You can turn left here and go through the placer tailings in Warren Meadows, about 0.5 mile, to a stream crossing. I parked here, as the water was high, but you can drive across later in the summer and continue down another 0.7 mile to the dredge remains at Site D. To reach Warren itself, from the bridge resume on the main road about 1.3 miles. The turn for Warren Creek at Charity Gulch is another 1.6 miles on the main road, or 2.9 miles from the bridge. This road continues on to Big Creek and also Stibnite, about 66 miles away. The road to Stibnite and Cinnabar is long and tiring, so be forewarned, but it is an amazing stretch of wild Idaho, and there is (usually) gas at Yellowpine.

Prospecting

Warren Creek, and particularly Warren Meadows, was an enormous producer in the 1860s. You can see white tailings piles over quite a distance, and they

These old dredge ruins are outside of Warren, and there are additional photography opportunities in the historic town

are primarily decomposing granite from the Idaho Batholith. Some of my sample pans had barely any black sand but still had color, which surprised me. Usually when you pan down and the last material all seems to wash away, there is nothing to see, but this was different. The spot on Steamboat Creek is at the foot of Halls Gulch, and there were at least three placer operations in this area. You can see extensive tailings; there are some lode mines and prospects up the road from that first bridge as you enter the district. The bridge locale listed here was good as well, and someone had created a small dam to run a sluice where Steamboat Creek meets Warren Creek. A handy access road is off to the left just past the bridge. At the crossing locale, I worked my way along a bedrock outcrop and liberated a few colors. You'll have to ford here to get to the dredge ruins. Warren itself gets busy during the summer, especially along the airstrip, and there is a US Forest Service facility in town. The Baum Shelter is a restaurant and bar that serves a mean hamburger and fries with "local" IPA on tap. The staff there might even direct you to a good panning spot. Parts of this town are quite photogenic and may look deserted, but you cannot explore this private property. Your best bet for exploring lode mines is to head up to the coordinates for the mouth of Charity Gulch. Note that there were multiple placers along this stretch of Warren Creek, so look for open access as well and see if the gold is more coarse than down at Warren Meadows. Turn right at the mouth of Charity Gulch and follow the tracks as far as you care to go. A rough 4WD road will connect you to Warren Summit and take you past several old prospects, or you can explore Keystone Meadows and Martinace Meadows.

83 Bear Valley

See map on page 175.
Land type: Creek
County: Valley
GPS:
A - Sign: 44.28373, -115.48390
B - Wash plant ruins: 44.28505, -115.48288
C - Pit: 44.28499, -115.48535
D - Sack Creek: 44.35895, -115.40809
E - Elk Creek: 44.41261, -115.37196
Best season: Late spring–summer
Land manager: Boise National Forest
Material: Very fine gold, rich black sands; garnets
Tools: Pan, sluice
Vehicle: Any; 4WD suggested—rough roads
Special attraction: Whitehawk Mountain lookout
Accommodations: Multiple developed campgrounds, such as Fir Creek, Elk Creek, and Bear Valley
Finding the sites: Reaching Bear Valley is easiest from Lowman on ID 21; just past the junction with CR 17, take Bear Valley Road north about 20 miles. There is an intersection with NF 569 headed left; this is Site A, with information signs about Bear Valley's black sands operation. From the sign, go right and follow the road north 0.1 mile to the concrete foundations of the old wash plant. This is Site B. Return to the signs and go northeast on NF 569 about 0.1 mile to the tracks that lead north to the remaining pits at Site C. This is the southern extent of the placer operation; you will see signs that workers rehabilitated the meadow, and several remaining ponds. The mouth of Sack Creek was also dredged, and there is a campground here; to reach it, drive north 7.8 miles on Bear Valley Road. To reach Site E and the campgrounds at Elk Creek, where Bear Valley Creek meets Elk Creek, drive north on Bear Valley Road for 4.8 miles.

Prospecting

Dredges in this area produced rare earth elements (REEs), also known as platinum group metals (PGMs). All of the creeks in this valley are full of black sands and a little gold, but it was the black sands that recent miners prized. When you concentrate on black sands, you can be a little less cautious, because you truly are going for quantity. Once you have enough to play with, use your magnet to separate the magnetite—you should be impressed by how much material avoids the magnet. You would need tons of this material to interest a smelter, but it still makes for an interesting addition to your collection of Idaho materials. The Sack Creek operation recovered niobium, columbium, zirconium, and more. The miners also reported uranium, zirconium, and titanium. Once you locate the foundations of the mill site at the southern end of the valley at Site B, you can locate access to the creek. There

Central Idaho East

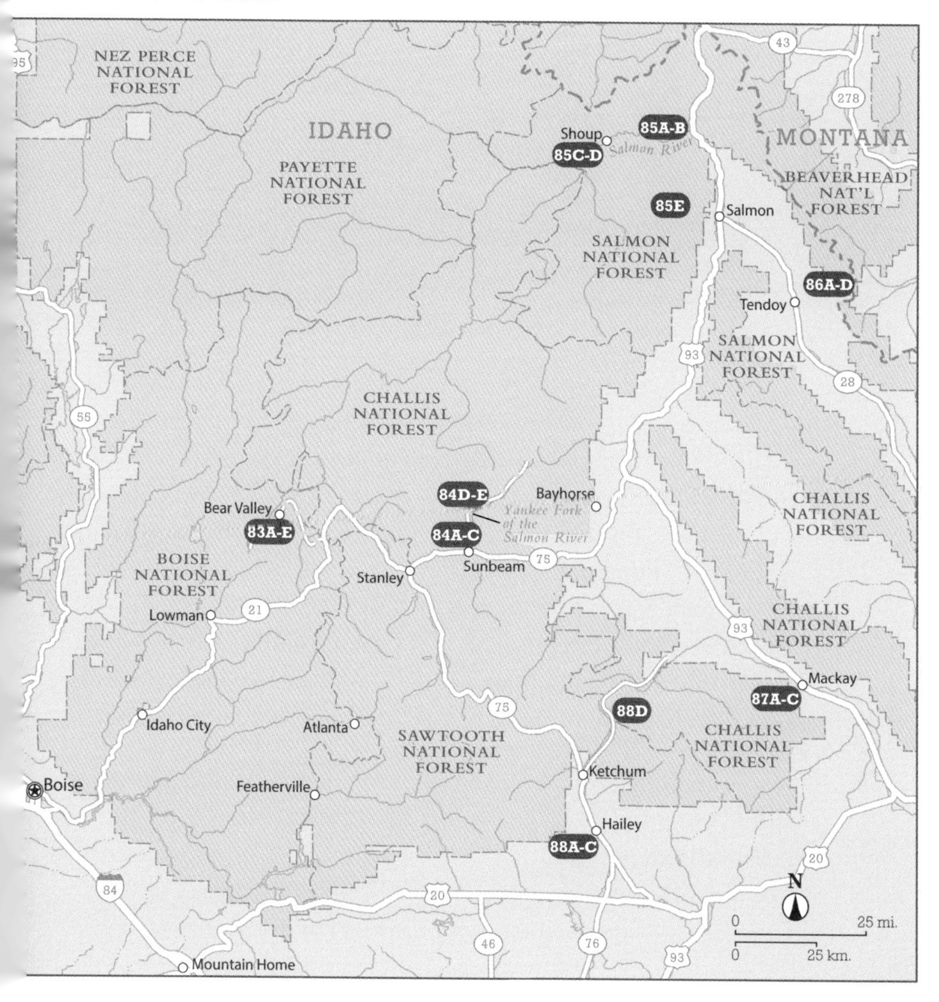

Bear Valley Creek winds through a wide valley before dumping a huge load of black sands into the Salmon River.

are a couple bridges at this southern end as well. The stretch between here and Sack Creek has numerous access points, each as good as the other. The bridge over the mouth of Sack Creek has good access and a nice bedrock exposure with cracks to at least provide a chance for gold. The Bear Valley campgrounds at Elk Creek offer multiple access points, with a nice beach at the oxbow and an even better beach right at the mouth of Bear Valley Creek. The Fir Creek Campground is better for camping than for panning and is only worthwhile by late summer, when the water is lowest. In case you are wondering, we have never seen any bears out here, but that doesn't mean they're not lurking.

84 Yankee Fork

See map on page 197.
Land type: Creek, tailings
County: Custer
GPS:
A - Sunbeam: 44.26972, -114.73445
B - Camp: 44.28058, -114.73276
C - Bridge: 44.28732, -114.72101
D - Dredge: 44.37704, -114.72292
E - Custer: 44.39077, -114.69219
Best season: Late summer
Land manager: Challis National Forest
Material: Fine gold, small flakes, black sands
Tools: Pan, sluice, highbanker, dredge, metal detector
Vehicle: Any; 4WD suggested for exploring
Special attractions: Yankee Fork dredge, Sunbeam Hot Springs
Accommodations: Numerous developed campgrounds on Yankee Fork and along Salmon River, such as Blind Rock and Flat Rock CGs; dispersed camping throughout the area
Finding the sites: The former community of Sunbeam is about 13.4 miles east of Stanley on ID 75. The mouth of the Yankee Fork is just east of the bridge—drive past it, about 0.1 mile total from the junction. Park safely past the guardrail and make your way down. A larger parking area is above the old breached Sunbeam Dam, but it is on the wrong side of Yankee Fork. Heading up Yankee Fork, there is a campground along the water; the turn is 0.7 mile from ID 75 on Yankee Fork Road. There is a good bridge-access spot 1 mile above the camp, and in another mile another campground, where the tailings start. The Yankee Fork dredge is about 8.4 miles from ID 75. To reach the old ghost town of Custer, continue on Yankee Fork Road by going right from the dredge about 2 miles; from the coordinates for Site E, you can see the remains of a mining structure to the south.

Prospecting

Check in at Sunbeam to learn more about the history of the district; the story behind the construction of a dam for dredge power is interesting. This

Volunteers restored the old dredge on the Yankee Fork of the Salmon River and opened it for tours.

district was a major producer in the late 1800s, with lode mines in the hills and dredges scouring the valley floor. Photographers enjoy the many rustic buildings and the Boot Hill cemetery above Bonanza, while prospectors have pulled color from just about any good access point. The tailings piles offer plenty of opportunities for metal detectors to look for big nuggets that got away from the dredge, and many of the tributaries that flow into the valley contain good gold close to their mouths. The Lucky Boy Mine area, on the southern side of Mount Bonanza at 44.3737, -114.6759, is accessible via Fourth of July Creek, about 1.2 miles beyond Custer, but could be gated. There are more dredge tailings on Jordan Creek near the dredge, but everything is claimed for several miles below the massive Jordan Mine, so I didn't include any coordinates along it. Maps also show multiple lode mines on Estes Mountain via Loon Creek Road. In fact, the number of old mines and prospects in this general vicinity is staggering. Bayhorse, another well-preserved mining town, is about 37 miles east of Sunbeam and also worth checking out.

85 Shoup

See map on page 197.
Land type: Creek, riverbank
County: Lemhi
GPS:
A - Buster Gulch: 45.37981, -114.06968
B - Pine Creek: 45.36334, -114.30056
C - Ebeneezer Placer: 45.303938, -114.51631
D - Golden Eagle Placer: 45.30141, -114.55167
E - Napias Creek: 45.16977, -114.15655
Best season: Late summer
Land manager: Salmon National Forest
Material: Fine gold, small flakes, black sands
Tools: Pan, hammer; dredge OK on Napias Creek
Vehicle: Any; 4WD suggested, required to reach Napias
Special attractions: Leesburg at 45.2239, -114.1133, Indian petroglyphs along the Salmon
Accommodations: Developed campgrounds at Ebeneezer and farther down the Salmon; dispersed camping up Panther Creek and Napias Creek
Finding the sites: The town of North Fork is on US 93, about 21.2 miles north of Salmon. From North Fork, drive west on NF 30/Salmon River Road about 4.8 miles. Watch for bighorn sheep coming down to the Salmon River for a drink of water as you drive, plus moose along the shore. Just before you get to Site A at Buster Gulch, there is a picnic area along the river, but the panning is not very good. The mines at Buster Gulch are behind a gate, so you cannot drive in. Look for great access to the river off to the left. To reach Site B at Pine Creek, drive past Indianola where Indian Creek comes in. If you have time, drive up Indian Creek to the old ghost town of Ulysses. Otherwise, keep going past Shoup (although this is a good spot for supplies). The bridge at Pine Creek is about 15 miles from Buster Gulch. To reach the old Ebeneezer Placer spot, keep going west on Salmon River Road to Ebeneezer Bar. You should see a sign for Indian petroglyphs at 45.3087, -114.5101, and then a camping area. The Golden Eagle Placer spot is about 2 miles farther west. The end of the road is another 10.4 miles and will take you to the Cunningham Bar, Proctor Bar, Kitchen Creek Bar, and Middle Fork Placers. Finally,

Watch for bighorn sheep on the road between Salmon and Shoup.

there is an interesting district near here, close to Leesburg, on Napias Creek. About 6.7 miles west of Pine Creek, there is a turnoff south onto Panther Creek Road. Travel 18.8 miles on this road to Napias Creek Road, turn left, and go 2.9 miles to where it connects to NF 242. After 1.5 miles you will see a turnoff down to the California Bar; this is Site E. The old ghost town of Leesburg is 4.7 miles farther north on NF 242.

Prospecting

The Salmon River is Wild and Scenic throughout this stretch, and for good reason. Moose, bighorn sheep, eagles, deer . . . it's an Idaho paradise. At Buster Gulch you can see across the river the remnants of a dam breech that occurred far upstream on Dump Creek. There were major placers farther up Dump Creek that you can reach via Leesburg, but at least there is good color at the Buster Gulch wayside. The old town of Ulysses, which sprang up around the Ulysses Mine, is about 5 miles off Salmon River Road up Indian Creek. Commodities from that district included selenium, bismuth, copper, lead, and zinc. You might be able to pick up some up-to-date local info at Shoup, so it is definitely worth a stop. Sometimes, even when I know where I'm going, I like to ask bartenders, cashiers, and gas station attendants if there's any good place to pan. Local intelligence is always best. Just past Shoup you can spot the Grunter Mine, which dates to 1882. There are

Farther down, there are some famous Indian petroglyphs.

numerous picturesque ruins, and the tailings are interesting. At Pine Creek, you can sample the Salmon, or you can work your way up Pine Creek and look for access to the creek. There were numerous placer operations along here, and lode mines too; the mouth of the creek has good color. Drive beyond the turn to Panther Creek to access the final two spots on the Salmon River. You'll pass numerous old placer diggings, including the Golden Queen, Upper Owl Creek, Lower Owl Creek, Homestake, Poverty Flat . . . anywhere the river formed a bend or a bar, the miners worked it. Beyond the Golden Eagle there are more old placer diggings all the way to the end of the road, and rafters could work even more with the right guide. Finally, Napias Creek is tough to reach, but it has good color, and nearby Mackinaw Creek also yields good color.

Leesburg is a fascinating glimpse of both an old mining town and a new, fully permitted and environmentally conscious mine at the same time. Leesburg is 33 miles from Salmon, so you do not have to backtrack if you want to take a different (though very rough) route. In addition, while you are in the area, Cobalt is about 13 miles south of Leesburg. However, the last time I was up here, the creek was reddish orange from the mines above Cobalt, and there was some major rehabilitation work going on, so there were gates on the roads to the mining district.

86 Lemhi Pass

See map on page 197.
Land type: Creek, mining ruins
County: Lemhi
GPS:
A - Agency Creek CG: 44.94994, -113.55802
B - Ruins: 44.99113, -113.49497
C - Agency Creek: 44.97294, -113.48226
D - Copper Queen Mine: 44.96939, -113.47578
Best season: Late spring–late summer
Land manager: Salmon National Forest
Material: Fine gold, small flakes, black sands
Tools: Pan, sluice, hammer
Vehicle: Any; 4WD suggested, required if leaving main road for Copper Queen and ruins
Special attractions: Lewis and Clark Trail, Lemhi Pass
Accommodations: Roadside park just out of Tendoy; Sacajawea Memorial CG at Lemhi Pass
Finding the sites: Locate Tendoy on ID 28, about 28 miles from US 93 southeast of Salmon. Take ID 28 South east from Tendoy, go about 0.2 mile, and turn right. Go another 0.2 mile and turn left onto Agency Creek Road. The roadside park for Site A is about 4.7 miles from this turn. To reach the old mining ruins under the power line and up the hill, drive from the roadside park about 4 miles toward the pass, then take a sharp left onto NF 6238. Go up the hill about 1.2 miles, then take a right turn onto the spur that dead-ends at the cliff. To reach the panning area on Agency Creek, return to the main road and drive 0.9 mile; you will see tailings along the creek. You are now just below access to the Copper Queen Mine and the mill; continue up Agency Creek Road 0.1 mile at most, then turn right onto NF 68. The mill is about 0.4 mile up this road, and more mine ruins are another 0.2 mile up.

Prospecting

Like Bear Valley, this district was also known for rare earth elements (REEs) in the black sands, but there is even more mineralization here. The Copper Queen Mine was a good copper producer, with silver and molybdenum by-products. Most of the roads above the Copper Queen lead to prospects, so if you have

Remnants of the mill near the Copper Queen Mine, just below Lemhi Pass

the right vehicle, you can have fun exploring the hills if the gates aren't in place. The mining ruins under the power lines have good tailings to explore, with reports of uranium and thorium in the black sands, so you probably do not want to keep your sample in your front pants pocket. If you have time, you can scoop up a couple buckets of pay dirt from the road where it crosses over the stream and work the buckets back at the Agency Creek Campground.

The hills above Agency Creek are dotted with prospects and a couple adits here and there. If you reach Lemhi Pass, which has a memorial to Sacajawea, look for the road that heads south, along the Continental Divide, follow it, and look for more outcrops to sample. Finally, at the Copper Queen ruins, you will find decent hand samples all over the place, mostly white quartz with rusty staining and green malachite. The ruins are hazardous and not worth exploring inside. There are multiple roads leading away from here, and they all seem to lead to prospects and mines that probably trucked ore down to the mill. The USGS topo map shows one adit northwest of the mill ruins. There was barely a trickle of water in the stream here, but again, you could bring pay dirt to Agency Creek Campground and process it there, and you can also pan colors from gravels at the campground.

87 Mackay

See map on page 197.
Land type: Mining ruins, hillside deposits
County: Custer
GPS:
A - Cossack Tunnel: 43.893166, -113.661311
B - Horseshoe Mine: 43.89858, -113.67909
C - Tram: 43.890232, -113.675386
Best season: Spring–summer
Land manager: Challis National Forest, BLM–Challis
Material: Sulfide ores
Tools: Pan, hammer, camera
Vehicle: Any; 4WD suggested
Special attraction: Mackay
Accommodations: Mackay Tourist Park, Park River RV Park; dispersed camping around Mackay Peak, White Knob Mountains (fire danger makes this difficult later in summer)
Finding the sites: Mackay is on US 93, about 26 miles north of Arco. There are brochures for the Mackay Mine Hill Tour in town wherever you see tourist information, or you can download it from the web at www.blm.gov/sites/default/files/docs/2021-07/BLM_ID_Mackay_Mine_Hill_Tour_Map_2021.pdf. The highlights for me are the mine, the ore zone, and the tram. Every time I have returned here, the routes and opportunities have changed, so I highly recommend downloading the current map. From US 93, turn southwest onto Seefried Lane and follow it for 1 mile, then continue onto White Knob Road for 2.4 miles. You won't see a tunnel; the mine buildings here mark Site A. Take NF 496 for 0.4 mile, then stay right for 0.3 mile to the multiple mines and prospects at the Horseshoe Mine (Site B). Return to NF 496, head up the hill to Smelter Road, and continue left to the coordinates for Site C at the Empire Mine ore bin.

Prospecting

The local preservation society has done an impressive job so far in restoring some of Mackay's mining glory. It was mainly a copper-mining camp, but many of the prospects held good gold values, along with zinc, silver,

The Mackay Mine Hill Tour is an awesome combination of history, preservation, and mining—go at your own pace, and take your time. The best tailings piles are farther up the mountain.

tungsten, molybdenum, and lead. These are classic skarn deposits, rather than quartz vein locales, and the ore can be unremarkable unless you bring your hand lens. The copper ores are the easiest by far to locate, as the classic green stain of malachite is unmistakable. The prettier blue azurite is much harder to find, but if you crack open enough stained rocks, you have a chance. Panning the river below the mining district brought a few colors back in the day, but in 2022 the park at 43.9027, -113.6169 prohibited panning. Try to budget a full day to explore the area, especially if you have a good vehicle. A careful minivan driver could probably make it to Site A, but these roads are tough on two-ply tires. If you enjoy these old mining ruins, consider a trip to Gilmore, Idaho, as well. See *Rockhounding Idaho* (FalconGuides, 2020) for more information.

88 Hailey

See map on page 197.
Land type: Creek, low hills, mountains
County: Blaine, Custer
GPS:
A - Bullion Street: 43.51679, -114.32209
B - Liberty Mine: 43.45652, -114.43839
C - Tip Top: 43.41548, -114.46348
D - Phi Kappa Mine: 43.84641, -114.20464
Best season: Late spring–summer
Land manager: Challis National Forest
Material: Fine gold, small flakes, black sands
Tools: Pan, hammer
Vehicle: Any; 4WD suggested near Hailey, required above Sun Valley
Special attraction: Hailey Hot Springs
Accommodations: Multiple campgrounds and RV parks in the Bellevue-Hailey-Ketchum area, such as Boundary and Park Creek above Sun Valley; dispersed camping (dry) way out Croy Creek; more dispersed camping up Trail Creek beyond the summit
Finding the sites: From Hailey on ID 75, Bullion Street turns into Croy Creek Road. The panning site at Bullion Street gets a bit crowded during the summer. The bridge is about 0.4 mile from the city center. To reach the turn for the Liberty Mine, head out Croy Creek Road about 7.2 miles farther and look for a slight right. It is just 0.3 mile from here, so it's a short walk. Bring a camera. To reach the Tip Top, go 4.9 miles farther on Croy Creek Road from the turn to the Liberty Mine, then look for a bad road leading up the hill. Go about 0.6 mile to the coordinates; it will be hard to tell this spot from the other diggings without a GPS, as there are large, gray rock dumps on both sides of the road here, and the Camas Mine is below you to the west. To reach the Phi Kappa Mine, head north on ID 75 about 12 miles to Ketchum and turn onto Sun Valley Road. Go 2.4 miles; the road is now NF 51/Trail Creek Road. Go 13.8 more miles, up over the top and into a nice little valley. Turn right onto Phi Kappa Creek Road and follow the road into the zigzags for about 1.3 miles.

Prospecting

The Hailey Gold Belt was a major producer in Idaho mining history; the hills west of Bellevue and Hailey are dotted with mines and prospects.

It's not hard to spot the mineralization out on Croy Creek, west of Hailey. Look for bright red and orange—it's as though the rocks were set on fire.

Unfortunately, there is not much water out here for panning. Croy Creek offers a good bet at its mouth with the Big Wood River—work your way upstream from the bridge. You will need a good 4WD to explore the hills in the gold area, or be willing to hike. Croy Creek Road itself is not too bad, but once you leave it, you will need power, a low gear, climbing ability, and not a little courage. Look for outcrops in this area that are a fiery red, either quartz or calcite, and sometimes "burnt" to yellow. You can collect a few buckets of this material, grind it up, and try panning it back at the water.

For additional ruins to explore, the road up to the Democrat Mine is decent until you have to dodge a giant cliff. The mine is at 43.5351, -114.3999. You can also swing a rock hammer at the Red Elephant Mine, an easier drive to 43.4894, -114.4315. Up at Phi Kappa, the altitude is close to 8,000 feet, but there is good quartz here, some with the rusty-orange staining prospectors always look for. If you can, find some sulfides by breaking up larger rocks—this material contained cadmium, copper, gold, lead, silver, and zinc, among other metals. The mine saw activity in the 1880s and has not seen much since then, but it gets you out of the dry hills around Hailey and is in nice country. Check out *Rockhounding Idaho* (FalconGuides, 2020) for information about nearby drusy quartz deposits and fossil graptolites on the drive up Trail Creek Road.

BOISE BASIN (IDAHO)

Boise Basin

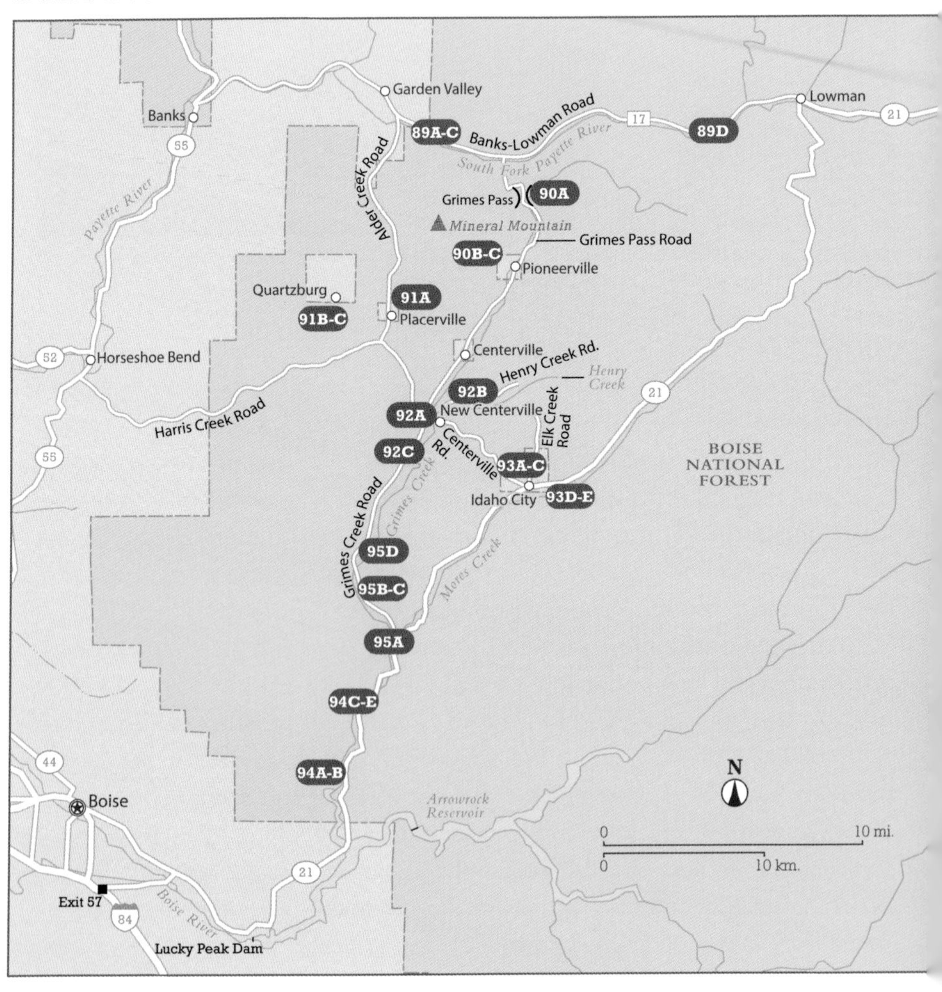

89 South Fork Payette River

Land type: Riverbank
County: Boise
GPS:
A - Alder Creek bridge: 44.07055, -115.94605
B - Hot Springs: 44.05391, -115.90808
C - Horseshoe: 44.05025, -115.88979
D - 60 Pine Flats CG: 44.06212, -115.68165
Best season: Late summer
Land manager: Boise National Forest
Material: Fine gold, small flakes, black sands
Tools: Pan, sluice, highbanker, dredge
Vehicle: Any
Special attraction: Lowman
Accommodations: 60 Pine Flats, Deadwood, and more developed campgrounds; limited dispersed camping on river
Finding the sites: Start from Banks on ID 55, about 42 miles north of Boise. Just north of Banks, turn east onto Banks-Lowman Road and drive 11.6 miles to the turn for Alder Creek Road. In 0.2 mile you will see parking that gives you access to the water at the bridge. The Hot Springs locale is 13.9 miles from Banks, and the Horseshoe locale is 14.9 miles from Banks. The next stretch of the river is quite steep, with limited access to anything resembling a bar. The turnoff to the campground at 60 Pine Flats is 28.1 miles from Banks, or just 5.8 miles from Lowman.

Prospecting

In Tom Bohmker's guide (2012) to Idaho gold prospecting, he notes that the South Fork of the Payette drains the northern edge of the Boise Basin district. The river was never as productive as Grimes Creek or Mores Creek, and the miners along the Payette were prone to head off to whatever new rush they got wind of. That meant the river never got diverted, ditched, flumed, or otherwise heavily worked, and remains attractive for panners to this day. The area around the Alder Creek bridge is a good warm-up if you plan to visit the Boise Basin, and there are some nice cracks in the bedrock here. At the Hot

Any one of these would work in the Payette's steep canyon; dredges up to 5 inches are legal, but these would be easier to work with in the steeper parts of the canyon.

Springs locale, there is a beach and easy access. There is more good access at the Horseshoe locale, but I would rather explore it from South Fork Road, which hugs the south shore; you can reach it after crossing the Alder Creek bridge. I have not checked there yet for good access. Finally, the 60 Pine Flats Campground is a good spot after all the steep cliffs. There was a placer operation at Lowman, even farther up the river, so don't despair that you have left the mineralized area at 60 Pine Flats.

90 Pioneerville

See map on page 210.
Land type: Creek
County: Boise
GPS:
A - Grimes Pass: 44.02037, -115.84423
B - Pioneerville: 43.96814, -115.84754
C - Upper Grimes Creek: 43.97126, -115.85035
Best season: Late summer
Land manager: BLM–Boise, Boise National Forest
Material: Fine gold, small flakes, black sands
Tools: Pan, bucket, hammer
Vehicle: Any; 4WD suggested for exploring
Special attraction: Grimes Monument
Accommodations: Dispersed camping above Pioneerville along Grimes Creek; Grayback CG near Idaho City and multiple more along the Payette River
Finding the sites: You can reach Pioneerville from the north via Grimes Pass by exiting ID 55 above Banks onto Banks-Lowman Road, then driving 11.6 miles to the Alder Creek Road turn. Go 0.3 mile on Alder Creek Road/NF 615, turn left onto NF 382, go 0.6 mile, then turn right onto NF 690. Drive 0.2 mile, stay left, and go another 2.4 miles. Turn left onto NF 652/Norn Creek Road, go 1 mile, then take the first left onto NF 382. Stay on NF 382 for 8.1 miles to reach the pass. To reach Pioneerville, continue on Grimes Pass south about 4.4 miles. To reach the Upper Grimes Creek site, backtrack slightly about 0.1 mile, take a left onto NF 395A, and drive about 0.2 mile to the creek.

Prospecting

The hills above Pioneerville are dotted with prospects and old mines, and it's worth exploring if you have a sturdy 4WD vehicle. The original name for this community was "Hogem," because the initial claim stakers hogged all the good ground. This area was a major supplier of the gold in Grimes Creek, and just about every road above the town or leading from the pass has a prospect or digging. The challenge is lots of private and patented land, which you can spot on the Boise National Forest map as white squares and rectangles among

At one time, Pioneerville boasted a population of 2,000 people. The old dance hall still stands.

the green US Forest Service land. For example, the Golden Age Mine was a big producer, but it is on private land. The Mountain Queen Mine is apparently abandoned and sits just west of Grimes Pass at 43.99127, -115.83679. There are several tailings piles throughout this area where you can get some hand samples of rusty-stained quartz. The lode mines up here tracked veins in quartz and calcite along with seams and fissures in rhyolite, granite, quartz monzonite, and diorite, which are common. Grimes Pass was where reports say Indians killed George Grimes, and his team hastily buried his remains. Others believe his greedy partners killed Grimes, but it is anyone's guess at this point. The spot on Grimes Creek listed here sits among piles of tailings, and the old-timers scoured it well. You should get colors, however. This road continues up Muddy Creek, and it is worth checking for open ground along there as well, especially above the tailings piles. There are lots of claims to dodge up here, too, so be aware.

91 Placerville

See map on page 210.
Land type: Creek
County: Boise
GPS:
A - Placerville Museum: 43.94306, -115.94566
B - Quartzburg: 43.94711, -115.98701
C - Belshazzar Mine: 43.94362, -116.01363
Best season: Spring–fall
Land manager: Boise National Forest, BLM–Boise
Material: Fine gold, small flakes, black sands
Tools: Pan, hammer
Vehicle: Any for Placerville; 4WD suggested for Quartzburg, required for Belshazzar
Special attractions: Placerville museums
Accommodations: Dispersed camping on USFS land at Alder Creek summit; Grayback CG near Idaho City
Finding the sites: If coming in from the north, turn east off ID 55 onto Banks-Lowman Road, drive 11.6 miles, and turn right (south) onto Alder Creek Road/NF 615. After 10.6 miles you will cross the summit and come into town. From Idaho City, drive north on Main Street to Centerville Road and turn left. Go 11.1 miles northwest, through the tailings fields, and in another 0.6 mile turn right onto Placerville Road/NF 615. Go 1.2 miles to reach the town. To access Quartzburg, drive west on Granite Street past the old cemetery; the road becomes NF 343/Granite Creek Road. Stay on this road for 2.7 miles, and then notice a nice clearing to the right, with a small creek down below. The main road to Quartzburg continues but dead-ends soon; the road to the Belshazzar site is on your left. It is rough and rutted, but after 1.9 miles you will see ruins.

Prospecting

The Placerville community is justifiably proud of their heritage, and it is very convenient to have a grocery store there. In the 1860s Placerville was the first camp in the Boise Basin for miners coming in from the north via the Payette River, and it grew to an estimated 3,000 souls by 1863. Today it has just a

Be sure to stock up on supplies at the Placerville store and also check on local conditions. Leave time for the museums in town.

handful of folks, but the town is pretty, with a sizable park in the center and a fascinating cemetery. If you come in via passenger car or minivan, you will still enjoy the visit, but you will need a sturdy vehicle to explore. It is bumpy, dusty, and potholed, with serious washboard sections once you leave the main roads. There is very little creek access that isn't claimed, so beware. This is close to the eastern edge of the Boise Basin mineralization belt, but the Gold Hill and Iowa Mines at Quartzburg were big producers back in the day. Now they are environmental eyesores under study by the State of Idaho for serious remediation. The roads in are gated.

There are quartz hand specimens throughout this area that represent the veins that miners worked. If you have a good vehicle with a paint job you don't mind scratching up, the road to the Belshazzar Mine starts right where you are parked. There was no gate the last time I checked, but if there is, it is a 2-mile hike (4 miles round-trip) and well worth the time if you can spare it. The last stretch before you reach the ruins parallels Fall Creek, which was a major producer. There are extensive tailings on Fall Creek down below by Mud Flat. Be sure to bring a good hammer to break apart the tailings.

92 New Centerville

See map on page 210.
Land type: Creek
County: Boise
GPS:
A - Bridge: 43.88284, -115.91314
B - Henry Creek: 43.88632, -115.88306
C - Grimes Creek: 43.86412, -115.91629
Best season: Spring–fall
Land manager: Boise National Forest, BLM–Boise
Material: Fine gold, small flakes, black sands
Tools: Pan, sluice, shovel, bucket
Vehicle: Any, but park carefully; 4WD suggested to explore.
Special attraction: Idaho City
Accommodations: Grayback CG; more campgrounds along ID 21 northeast of Idaho City; good dispersed camping along Henry Creek
Finding the sites: From Idaho City drive north on Main Street to Centerville Road and turn left. Drive up and over the hill 7.2 miles to the bridge across Grimes Creek. From this bridge, backtrack toward Idaho City about 0.2 mile to Henry Creek Road/NF 399 and turn left. Stay on this "main" Henry Creek Road about 1.7 miles. To reach the other Grimes Creek location, go back to Centerville Road and make a slight jog left then right to Grimes Creek Road/NF 364, then drive about 1.3 miles.

Prospecting

New Centerville is basically the intersection of Grimes Creek and Centerville Road, and it is the heart of the Boise Basin. "Old" Centerville is devoid of any historic buildings, but at one time it was a handsome community of 3,000, with daily stages to Placerville, Pioneerville, and Idaho City. Today it sits among massive tailings piles, with frequent ponds where the dredges sat. Water was always a concern as the hot summer wore on for the miners. The bridge locale at Site A is the easiest to access, with parking on the northwest end of the bridge. Up Henry Creek at Site B you will find yet more tailings and less water, but there are good colors and access to a bewildering network

The flats near Centerville have been hydraulicked, dredged, and more. This aging wash plant is close to New Centerville.

of roads. The Grimes Creek site southwest of the bridge remains unclaimed as of 2022, so probably sits on old tailings. You won't have much hope of reaching bedrock out here, and there aren't even good boulder piles to work around, but if you can locate even a small bend in the creeks, you have a pretty good shot at pulling your own Boise Basin sample. Reports from the early days up here are sketchy, but an estimated $1 billion worth of gold came from the entire district. There are multiple operations here to this day, and the tailings piles stretch for miles. The entire drive from the bridge, past "Old" Centerville to Pioneerville, crosses private land, but the locales here are open, mostly because the easiest gold left long ago.

93 Idaho City

See map on page 210.

Land type: City, creek, hillside

County: Boise

GPS:

A - Visitor center: 43.82571, -115.83296

B - Museum: 43.82847, -115.83372

C - Gold Hill: 43.83544, -115.82652

D - Steamboat Gulch: 43.82279, -115.80933

E - ATV park: 43.83194, -115.79084

Best season: Spring–fall

Land manager: Boise National Forest

Material: Fine gold, small flakes, black sands

Tools: Pan, sluice, shovel, bucket

Vehicle: Any; Gold Hill might be a little rough.

Special attraction: Diamond Lil's Saloon

Accommodations: Grayback CG; dispersed camping along Grimes Creek, up Mores Creek

Finding the sites: Idaho City is about 34 miles via ID 21 from the east edge of Boise. Leave I-84 at exit 57, drive past Lucky Peak Dam, and head straight for the visitor center. From there you can walk or drive about 5 blocks to the museum, located at Wall and Montgomery Streets. To view the old hydraulic workings at Gold Hill, leave the visitor center and drive up Main Street 0.7 mile; it will become Elk Creek Road. Turn right onto a faint dirt road and drive 0.3 mile until you see the impressive white cliffs where the "water cannons," or monitors, washed away the hillside. This was the Idaho City Placer Mine. To reach Mores Creek at Steamboat Gulch, travel east on ID 21 from the visitor center about 0.9 mile and turn right onto Steamboat Road. After just 0.3 mile you will see a small bridge, and you can park safely anywhere, or take a faint road to the left just before the bridge. The turn for the ATV park along Mores Creek is about 2.2 miles east of the visitor center on ID 21. Take the first left into the parking area and drive as close as you care to the creek. I like the upper northeast end.

The Boise Basin Museum in Idaho City offers good information on local conditions as well as excellent exhibits about mining in this famous district.

Prospecting

Every dedicated gold prospector in the Pacific Northwest should visit Idaho City at least once. The visitor center is helpful, the museum is fantastic, and Diamond Lil's is the kind of historic saloon that will put a smile on your face just walking in. I like the jaunt up to Gold Hill just to see how serious the old-timers were about using giant jets of water to wash down hillsides. The gold had to be good to put out that kind of effort, and it was. Mores Creek gets its name from J. Marion More, a politician of the day who reported on the early success of the miners. He told officials some miners were recovering $300 per day—back when gold was less than $20 per troy ounce. As you can imagine, the easiest gold is gone, and miners worked the best ground more than once. Some hot spots still host claims. However, there are places around Idaho City that you can still work, and even some of the gulches and tributary streams still yield color. The Steamboat area is a winter playground, but there is ample water in summer months—probably not enough to warrant a dredge, but a highbanker might work. At the ATV park you will find lots of activity during the summer, but it is not too hard to find a quiet spot where you can test a few pans or even set up something more serious. There are dozens of mines and prospects northeast of Idaho City along ID 21 all the way to Mores Creek summit, but you would need the right vehicle to explore those back roads, and you'd likely run into claims.

94 Mores Creek

See map on page 210.
Land type: Creek
County: Boise
GPS:
A - Robie Creek boat ramp: 43.62842, -115.99489
B - Mores Creek Park: 43.63922, -115.99364
C - Hummingbird: 43.65405, -115.97948
D - Rush Creek: 43.66452, -115.97892
E - Daggett Creek: 43.67807, -115.97308
Best season: Late summer
Land manager: Boise National Forest, BLM–Boise
Material: Fine gold, small flakes, black sands
Tools: Pan, sluice, shovel, bucket
Vehicle: Any
Special attraction: Lucky Peak Dam
Accommodations: Grayback CG; dispersed camping along Grimes Creek
Finding the sites: These sites are all along ID 21 or very near it, between Lucky Peak Dam and the junction with Grimes Creek. Start the drive from the turn to the viewpoint just past the top of the dam, about 6.5 miles from exit 57 on I-84 at Boise. Drive north on ID 21 for 10.7 miles to Robie Creek Road and turn left. Now backtrack on the other side of Mores Creek for 1.1 miles to Mores Creek Park and 2.1 miles to Robie Creek Park. The next spot is a wide pull-out on a good bend only 0.3 mile farther north on ID 21, just past Hummingbird Haven Road. There is another pull-out before you get to Rush Creek Road, just 0.8 mile farther up the road. This one is frequently dammed up by the locals to make an excellent swimming area. The final area is another 1.2 miles farther, on a big bend where Daggett Creek enters Mores Creek. This spot is 14.8 miles south of Idaho City.

Prospecting

Mores Creek contains good gold in this stretch, thanks to supply by Grimes Creek. This is the "drain" for the Boise Basin, in effect, and miners have worked this stretch since the discovery in the 1860s. Still, each spring flood moves the gravels around, and there is good color here. The bad news is claim

Giant "water cannons" like this one helped erode the cliffs along Mores Creek near Idaho City.

markers and private property, so be aware that these coordinates aren't guaranteed, except for the two parks at Robie Creek and Mores Creek, which are withdrawn from mineral entry. Being so close to Boise, this area is popular with weekend and summer tourists looking for a cooldown, and the lower spots tend to fill up. I liked Daggett Creek near the bridge, as the creek has a good bend there. I have seen a dredge at Rush Creek in the distant past, as it is a good pool. Robie Creek Park is underwater for parts of the year, and it can be a mud bath before it dries out. By late August you can sometimes see lines of black sands created by wave action as the water drew down.

95 Grimes Creek

See map on page 210.
Land type: Creek
County: Boise
GPS:
A - Bridge: 43.72676, -115.95301
B - Large bend: 43.74635, -115.96781
C - GPAA claim: 43.75213, -115.97519
D - IGPA claim: 43.78297, -115.97069
Best season: Late summer
Land manager: Boise National Forest, BLM–Boise
Material: Fine gold, small flakes, black sands
Tools: Pan, sluice, highbanker, dredge
Vehicle: Any; 4WD suggested the farther up you go
Special attraction: Idaho City
Accommodations: Camping all along Grimes Creek; developed campground at Grayback CG
Finding the sites: The mouth of Grimes Creek is on ID 21, at the bridge exactly 10 miles south of the Idaho City Visitor Center and about 24 miles from exit 57 on I-84 at Boise. The second site, a set of nice bends on Grimes Creek, is about 1.8 miles up Grimes Creek Road/NF 364. The old GPAA claim near Clear Creek Road is another 0.8 mile up the road; there is access off Grimes Creek Road and also from Clear Creek Road before the bridge. The Idaho Gold Prospectors Association (IGPA) claim is about 2.4 miles farther up Grimes Creek Road from the junction with Clear Creek Road.

Prospecting

Grimes Creek was a major producer from the beginning of the development of the Boise Basin. The two spots at the lower end of the creek are usually crowded with vacationers during the hot summer months, especially on weekends and particularly on major holiday weekends such as Memorial Day, Fourth of July, and Labor Day. In 2022 there was so much overuse by visitors that local authorities closed off some of the area around Labor Day. The two club claims are also popular and might convince you of the value of joining

Grimes Creek was the richest drainage in the Boise Basin.

a club. One problem out here during fire season is that you need to have an approved fire ring for campfires, which you will find in only a few camping areas along Grimes Creek—usually the ones that have vault toilets. There is not much shade out here until you get farther up in the canyon area above Macks Creek, so consider that as well. The gold is not consistent, and you will want to use all your best practices to get on an inside bend, around large boulders, and/or dig a deep hole. You will have a better chance at reaching bedrock farther up by the IGPA claim, and that might be a consideration if you are "shopping" for a club to join. Dredgers do well on Grimes Creek to this day, which is why so much of it is claimed or private. I have seen every kind of contraption out here, plus metal detector enthusiasts checking the tailings piles.

96 Atlanta

See map on page 210.
Land type: River, mine dumps
County: Elmore
GPS:
A - Middle Fork 1: 43.77533, -115.51791
B - Middle Fork 2: 43.78353, -115.46429
C - Atlanta Hot Springs: 43.81195, -115.11068
D - Mines: 43.78711, -115.10932
Best season: Late summer to dodge snow
Land manager: Boise National Forest
Material: Fine gold, small flakes, black sands
Tools: Pan, hammer
Vehicle: 4WD required—road is rough and long.
Special attractions: Atlanta Hot Springs, Chattanooga Hot Springs, and many more
Accommodations: Queens River, Riverside, and Power Plant CGs, among others; dispersed camping along Middle Fork of the Boise River
Finding the sites: Atlanta is at the end of a long, dusty, rough road no matter how you do it. For this guide, start at Idaho City and drive 2.2 miles on ID 21 East, then take Rabbit Creek Road/ NF 376 for 27 miles to CR 82/ Middle Fork Boise Road. Drive about 0.1 mile east, up the river, to the tracks that lead to the gravel bar. This is Site A. Site B is 3.2 miles farther east, and there are no active claims between these two readings. To reach Site C, resume travel east for 24 miles. Drive through Atlanta on Main Street/Middle Fork Boise River Road about 1.3 miles to the coordinates, just past the pond in a fenced area. To reach Site D, backtrack 0.9 mile, turn left, and drive 1.6 miles up the hill on Mine Hill Road. The mines are off-limits, but occasionally photogenic, and there are quartz samples in the road. Be cautious during working hours, as these mines are active.

Prospecting

This area is remote, but it was seeing more interest when Toronto-based Atlanta Gold Inc. lined up permits to begin a large open-pit operation. However, that interest seems to have faded; their atgoldinc.com website hasn't been updated since 2017. Prospector John Stanley and his party made the initial Atlanta discoveries in the summer of 1863 but kept the locations secret. Due to the extremely remote location, significant production started slowly after the initial ore discoveries in 1863, and it took almost seventy years for the district's potential to be fully developed. Today, eventual production from Atlanta-area mines is estimated at over $16 billion. Atlanta was a bustling little town back in the 1860s, and the Atlanta Lode has kept Atlanta alive ever since. Engineers measure the values today in grams per ton, thus the need for cyanide heap leaching and difficult environmental permitting. The Middle

Atlanta and Featherville

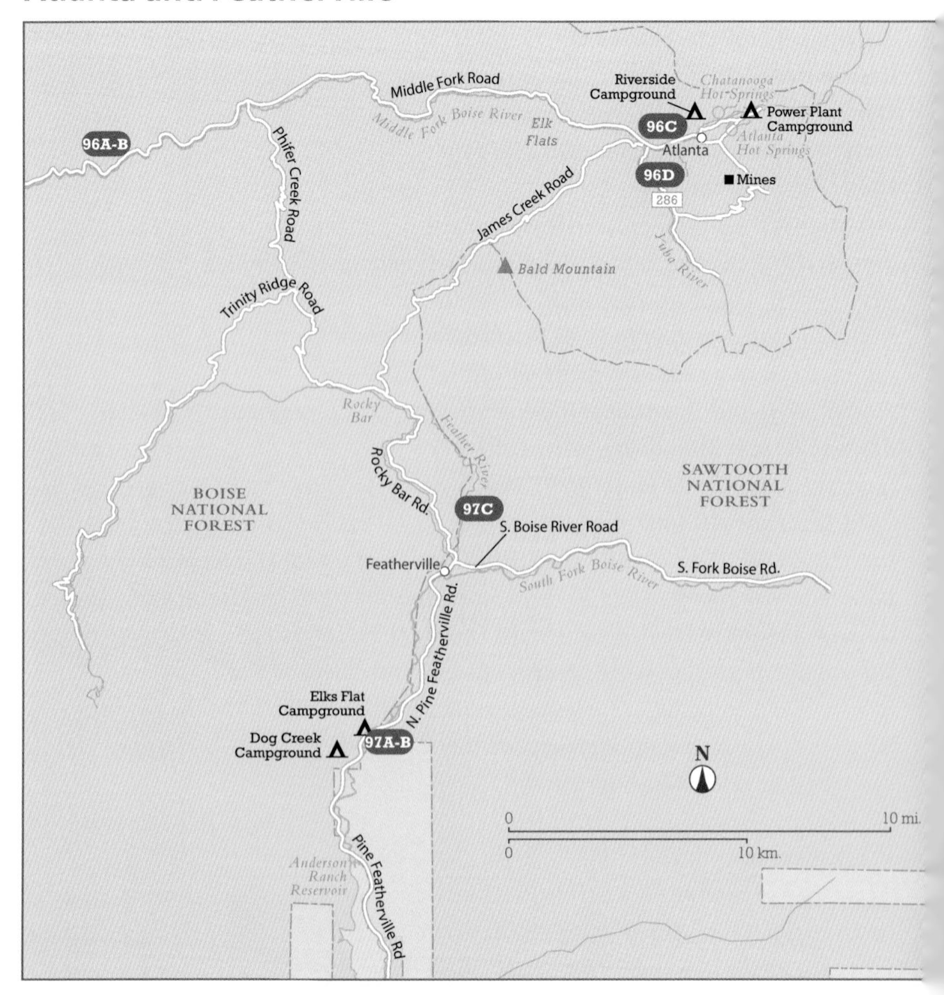

Atlanta Hot Springs is a welcome stop on any long gold trip.

Fork saw considerable dredging and still yields good color; finding an inside bend and bedrock that aren't under claim are the biggest challenges. Miners early on recognized that this was not an area for poor men; they would need to build a flume, ditch, or otherwise divert the water. Miners did well at the mouth of the Yuba, but attention soon focused on the silver and gold ores of the Atlanta Lode, and placer mining faded.

If you like to combine natural hot springs with gold prospecting, this is a great area, with dozens of warm and hot springs lining the Middle Fork, and excellent pools up at Atlanta. *Hiking Hot Springs in the Pacific Northwest*, by Evie Litton (FalconGuides, 2001), is an excellent guidebook covering this area.

97 Featherville

See map on page 226.

Land type: Creek

County: Elmore

GPS:

A - Dog Creek: 43.52982, -115.30085

B - Elk Flats CG, Hot Springs: 43.53911, -115.29064

C - Feather River: 43.63607, -115.24855

Best season: Late summer

Land manager: Boise National Forest

Material: Fine gold, small flakes, black sands

Tools: Pan, sluice, shovel, bucket

Vehicle: Any for lower spots; 4WD required above Featherville

Special attractions: Dismal Swamp, Paradise Hot Springs

Accommodations: Dog Creek and Elk Flats CGs, among others; dispersed camping above Featherville; lodges such as Paradise Hot Springs

Finding the sites: Featherville is about 23 miles south of Atlanta—a trip that can easily take an hour and a half. To reach Featherville from Boise, take exit 95 from I-84 and head 31.7 miles toward Fairfield on US 20/ID 51. Turn left onto CR 61/Louse Creek Road and drive 18.1 miles, past Anderson Ranch Reservoir, to Pine. Turn right onto Wood Creek Road/N Pine-Featherville Road and stay on the west side of the South Fork of the Boise River for about 3.4 miles until you see the turn for Dog Creek Campground. There is a pull-out along the river about 0.1 mile north from the Dog Creek Campground turn. The turn into the Elk Flats Campground is another 1 mile north on CR 61, and there is a nice beach on the other side of the bridge as well. To reach the upper spot, continue north on CR 61/Pine-Featherville Road toward Featherville, about 6.2 miles, and turn right onto South Boise River Road. Cross over the tailings below that congest the Feather River and take an immediate left onto Cayuse Creek Road. About 1.7 miles up there is access to the water before the road swings east up Cayuse Creek, and there is more access to the Feather River as you keep going north on an increasingly poor road.

The bridge near Paradise features great access to the South Fork of the Boise River from both banks. The hot springs are along that far bank.

Prospecting

Originally known as Junction Bar, Featherville marks where the Feather River meets the South Fork of the Boise River. Above here on the Feather River, some notable camps sprang up in the 1860s, including Happy Camp, Rocky Bar, and Spanish Town. I did not list them here because I have never had much luck at Rocky Bar finding open access. Placering was the primary source of gold in this area, as most lode deposits were spotty surface ledges that did not hold their value at depth. The GPAA has had a claim for years at Dog Creek at Site A, but its Thunder Dog claim does not include the mouth of Dog Creek, so stay north of the creek. The club also carries a claim 19 miles east on the South Fork, but I did not have much luck there either. At Elk Flats Campground there is good access, and in 2022 there appeared to be construction on new campgrounds on the opposite bank of the river. Note that there are shallow dug-out hot springs at 43.5398, -115.2891, on the east side of the river at the bridge. I like the Feather River spot at Site C because it is above where the dredges worked extensively. It has been open in the past, but be on the lookout for updated claim markers.

WESTERN IDAHO

98 Silver City

Land type: Creek, mountains
County: Owyhee
GPS:
A - Ruby City: 43.02607, -116.73380
B - Dewey Mine: 43.03984, -116.76577
C - Delamar Placer: 43.02479, -116.83864
D - Morningstar Mine: 43.01986, -116.73171
E - Campground: 43.01536, -116.73103
F - Trade Dollar: 43.00958, -116.74447
G - Golden Chariot: 43.00681, -116.69445
Best season: Summer
Land manager: BLM–Owyhee
Material: Fine gold, small flakes, black sands; cassiterite, quartz

Silver City

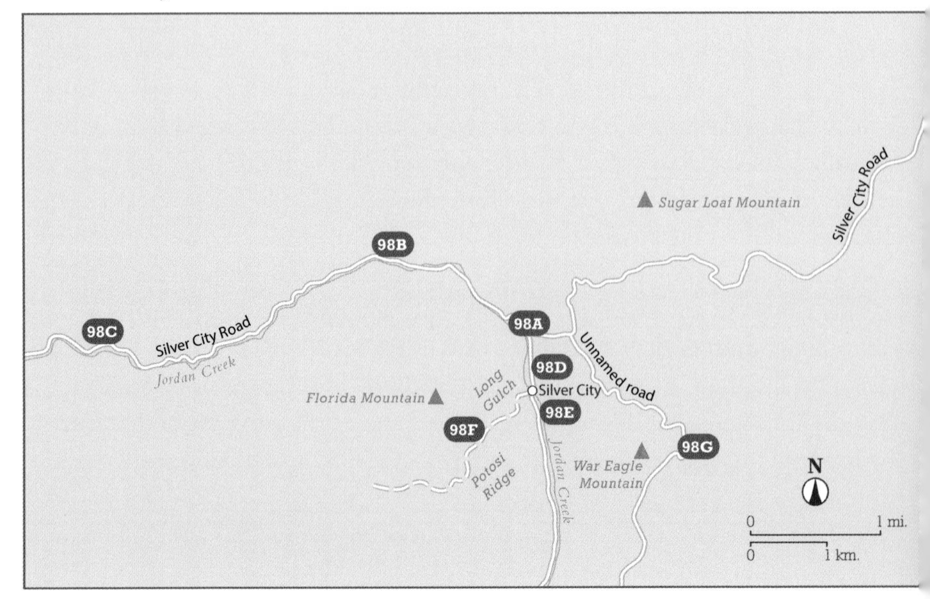

Famed Idaho Hotel in downtown Silver City offers food, drinks, and lodging. A gunfight took place here between competing mine owners.

Tools: Pan, sluice, highbanker, dredge
Vehicle: 4WD required—long, bumpy road with no gas at Silver City
Special attraction: Idaho Hotel
Accommodations: Campground at Silver City and Ruby City; dispersed camping throughout BLM lands
Finding the sites: Silver City is accessible via Jordan Valley, Oregon, on US 95; go east from Jordan Valley on Yturri Boulevard about 3.7 miles, where the road becomes Trout Creek Road, and go another 12.3 miles. Now turn left onto Silver City Road and drive 8.9 miles east and then due south to the town. Or come in from Nampa by driving south on ID 45 for 14.5 miles and across the Snake River to ID 78, then 14.3 miles south to Silver City Road. Turn right and drive 19.2 miles. From the Idaho Hotel, drive north on Washington Street 0.7 mile to the intersection with Silver City Road. This is the old Ruby City site; there is a primitive bathroom, and it's OK to camp here. These were the Good Hope Placers on Jordan Creek. Turn left and drive 2.3 miles east to the old Dewey Mine; you can also pan

here, or drive another 4.9 miles to the Delamar Placer diggings. The Morningstar Mine is at the end of Morningstar Road in Silver City, but be cautious if there have been too many careless visitors and new signs sprout up. To reach the Trade Dollar, leave Jordan Street for Washington Street at the Idaho Hotel, bear right, and follow the main road about 1.1 miles. The Golden Chariot site is a big intersection beneath War Eagle Mountain; there are multiple spots to check here. From Ruby City drive east 0.5 mile and take the first right up to War Eagle. Go about 2.8 miles; if there is activity at the War Eagle Mine, this road may be blocked, so be wary.

Prospecting

This area has a rich mining history, and while there was silver in the ore, a lot of gold came out of these hills. There was even a famous gunfight between two mine owners in front of the Idaho Hotel. Your first order of business upon arriving in Silver City is to go into the hotel and buy something—they sell an excellent oversize map of the area, and the owner has a ton of knowledge as well. If there have been any changes, such as new activity at the War Eagle Mine or a new reclamation project somewhere, they will know here. There is an excellent campground at the end of Jordan Street, and you can pan there or wander the hills and inspect a couple of prospects nearby. Rockhounds have long scoured these hills for agate, jasper, and petrified wood, especially down below Delamar. The mine dumps around Silver City contain gold, native silver, and quartz crystals, especially at the Golden Chariot and the Morningstar. Jordan Creek yields rounded lumps of tin ore (cassiterite) near Ruby City. There's a lot to explore out here, so leave yourself plenty of time. Claims come and go up here, so be aware that you may have to give up on a promising site due to changing local conditions.

99 Mineral

Land type: Creek, tailings, and dumps
County: Washington
GPS:
A - Boone Mine: 44.56842, -117.07698
B - Condor Mine: 44.56362, -117.06249
C - Mariah Mine: 44.56191, -117.06919
D - Open pit: 44.56899, -117.03709
Best season: Late spring–fall
Land manager: BLM–Boise, Payette National Forest
Material: Metal ores, including gold, silver, zinc, lead, antimony, and copper
Tools: Hammer; pan in Dennett Creek if you find water
Vehicle: 4WD required
Special attractions: Hells Canyon, Seven Devils
Accommodations: Semi-developed campgrounds along upper Mann Creek; dispersed camping at Mineral

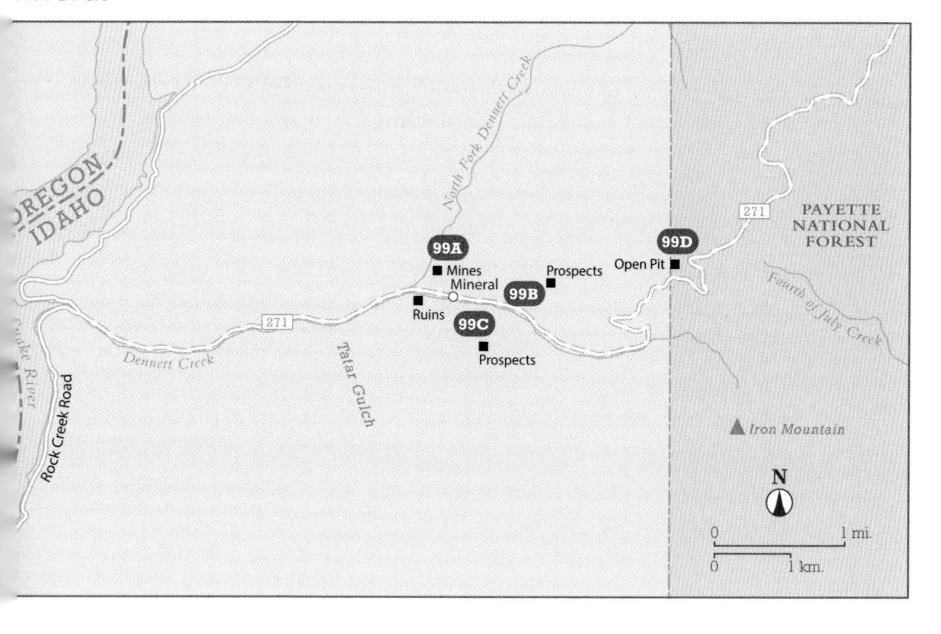

The Mineral mining district requires a long drive, but it isn't too hard to find—look for scrap metal, tailings, and old buildings.

Finding the sites: Mineral is a tough drive from Weiser, no matter which way you try. We took the back way in by hugging the Snake River on Rock Creek Road the first time, but it seemed to take forever. To take this route, leave Weiser by going northwest on Main Street to the US 95 South Spur, but instead of crossing the water, turn left onto CR 70. Take this road 3.8 miles to turn right onto Jonathan Road, and drive 2 miles to where it turns into Henley Basin/Rock Creek Road. Drive 12.6 miles, whereupon the main road turns left and is now Mineral Road. After 8.2 miles turn left onto Rock Creek Road, drive 5 miles, and turn right onto NF 271/ Dennett Creek. After 2.9 miles you will reach Mineral. At the old buildings along the road at Mineral, drive north on the dirt track 0.2 mile to see the ruins of the Boone Mine. There are numerous adits, ruins, and tailings piles on this hillside. Drive 0.7 mile east up Dennett Creek and look for the remains of the Condor Mine on the north side of the valley. Look for a road leading at an angle on the south side, which heads to the Mariah Mine. Another 2.8 miles up NF 271/Dennett Creek, you will see the open-pit mine at the intersection.

Prospecting

I only panned the slightest tiny colors out of Dennett Creek. The North Fork, which leads up to the bottom of the Boone Mine, had even less water. We camped right at the junction of the two creeks; there was plenty of wood. The entire hillside around the Boone Mine is fun to explore, and you should get some nice sulfide samples. Look for galena, sphalerite, malachite, pyrite, and many other minerals. The readings for the Condor Mine are a guess; the adits are either caved in or I didn't find them. There are places to grab samples in the bedrock on both sides of the valley, but it's a 500-foot climb either way. At least the walk to the Mariah isn't a complete bushwhack. The open pit is the easiest place to grab samples, as the road takes you almost right to the pit walls.

HONORABLE MENTIONS

The following Idaho locales did not make it into the book, but they're worth a visit if you're in the area.

T. Caribou City

This district, on Caribou Mountain in Bonneville County, is located in the far southeast corner of Idaho and is a district all by itself. It is sometimes referred to as Cariboo City, as it was named after "Cariboo Jack," a miner from the famed rush at Cariboo in British Columbia. I have been there and panned color in Iowa Creek, but I cannot vouch for much more than that until I get back there.

U. Chloride Gulch

I had more luck rockhounding up here than gold panning. This district is up by Pend Oreille Lake in the Lakeview District and has seen recent gold mining by modern methods. I panned some color out of the gulch, but I'd need to explore more of the territory up there to give it more weight.

V. Pearl

Located east of Emmett, this district is too dry to pan and too locked up with private property and claims to recommend. There's some good rockhounding on the tailings, as this area is more famous for Willow Creek picture jasper than gold.

W. Priest Lake

Way far up on the west side of Priest Lake, there are several old prospects and mines, particularly the Continental Mine. We panned tiny flour gold at Gold Creek and found a single quartz crystal on Quartz Crystals Road, but the area is too remote and not very abundant.

X. Seven Devils

Probably one of my favorite rockhounding districts in Idaho and covered in my FalconGuide *Rockhounding Idaho*. Unfortunately, the gold is very limited. There used to be some placer diggings near Bear, but they're on private land now. The copper mines above Cuprum are full of malachite, azurite, bornite, and other sulfides, but they are very limited in gold, so I left them out.

Y. Snake River Placers

If you don't mind playing with impossibly fine flour gold, you can prospect for gold all along the Snake River. From American Falls to the mouth of the Boise River, the Snake looks like it could be productive if you could move sand by the ton. Personally, I find the Snake to be somewhat smelly by June, as it drains hundreds of thousands of acres of farmland, feedlots, and dairy operations. Combine that with the impossibly fine gold, and I left it out. It's true that there was once a productive placer in Hells Canyon operated by Chinese miners. The book *Massacred for Gold* by R. Gregory Nokes details what happened when a band of marauding cowboys descended on the miners, robbed and killed them, and tossed the bodies into the river.

Z. Stibnite

The area around Yellow Pine, Stibnite, and Cinnabar hosts great mineralization, but there is only limited gold in most of the streams in this area. There's great history out here, and the scenery is terrific, but it's not worth mounting a panning expedition here.

The giant cut at Stibnite yielded mostly base metals.

APPENDIX A: MODERN TOOLS

In my companion book, *The Modern Rockhounding and Prospecting Handbook*, available through FalconGuides, you'll learn the basics of reading geology maps, understanding symbols, figuring out basic geology, and more. Here you'll get overview information about various types of machines and devices, and you can form your own opinion.

Buckets, Shovels, and Screens

I've met quite a few prospectors who enjoy bringing home a nice bucket of concentrates and then pan in the warmth of their garage. They don't like stretching out into a creek or river, trying to find a deep spot, or risking a bad back, and they don't like getting their clothes and shoes wet when they

Screening material is an important step in getting good pay dirt to turn into concentrates.

inevitably slip. At home you can use warm water, stay out of the wind, put your panning tub up on a bench, add a drop of dispersant to cut down on surface tension, and bring in plenty of light. So the first tools you need out there are a bucket, a shovel, and a screen. And truthfully, the screens are optional but a good idea, so you aren't bringing home a bunch of big rocks that you don't really need. Because the odds are billions and billions to one that you'll ever screen a big nugget, you might as well use screens to improve the quality of your pay dirt—whether you pan it out in the field or bring it home.

Shovels, buckets, and screens do not require a dredging permit, and they're quiet. Your goal is to dig a big hole, because the deeper you go, the bigger the gold in 99 percent of the deposits. So it's up to you if you bring the same shovel that you use in the garden or get a shorter shovel with a nice handle. On some trips you may use a trowel more than any other shovel, especially if you are digging out crevices in bedrock. The more you dig, and the deeper you go, the better your chances, so these are your first tools to consider.

Gold Pans

Your next tool is the venerable gold pan. Some older folks skip this step and go straight to spiral pans, but every prospector should have a pan or two in his or her collection.

The only way to get good at gold prospecting is through practice—you have to pan at least a dozen pans before you start to feel comfortable with sliding the heavies around, breaking up the muck, and trusting the riffles. There are twenty or more modern pan designs today, with sharp riffles, double sets of riffles for rough and fine panning, and all kinds of interesting shapes and additions. My personal favorite right now is a 14-inch octagonal pan with two sets of riffles and a broad, flat bottom. What this pan helps with is avoiding a tendency to swirl material in a circular pattern, which only serves to spread the concentrates around on the bottom. What you want to do is continually push heavy material into the first riffle and lock it in there, then you can start to feel more nonchalant about everything else in the pan. When you are first breaking down a big pan full of muck, you need to learn to shake the pan vigorously so that the heavy material at the bottom slides forward to that first notch. This is called "stratifying," and it only works when you get the muck entirely into solution so that it's behaving like a liquid. Once you get your material into suspension, gravity will pull things apart, with light material like clay and debris floating to the top, and heavy material, like gold, garnets, and

Close-up showing the two sets of riffles on the 14-inch Octapan

black sands, sinking to the bottom. That means less swirling as you get closer to the end, and more side-to-side action. Pack that first riffle, and trust science.

The great thing about panning is that it's easy, and it's almost never banned. Even Wild and Scenic Rivers are open to panning below the high-water line. National parks and monuments are not open, however. Some state parks also restrict panning; but county parks, US Forest Service campgrounds, and BLM campgrounds are usually open. So it's the first skill you need to gain as a prospector.

Snuffer Bottles

Once you've panned down a pan to where you can see what you have, you need some way to save the sample. You can use a wide-mouthed glass jar and just dump your concentrates in, although it's a good idea to have a second pan underneath the jar. Most modern prospectors use a plastic "snuffer bottle," which is a small plastic bottle with a removable straw. Once you're good at it, you can slurp up your black sands quickly. There are two modes—without the straw, for bulk recovery, and with the straw, for finer work. Once you get good with a snuffer bottle, you can make your specimen cleaner by picking up just the black sands and leaving the gold, for example—if you have plenty

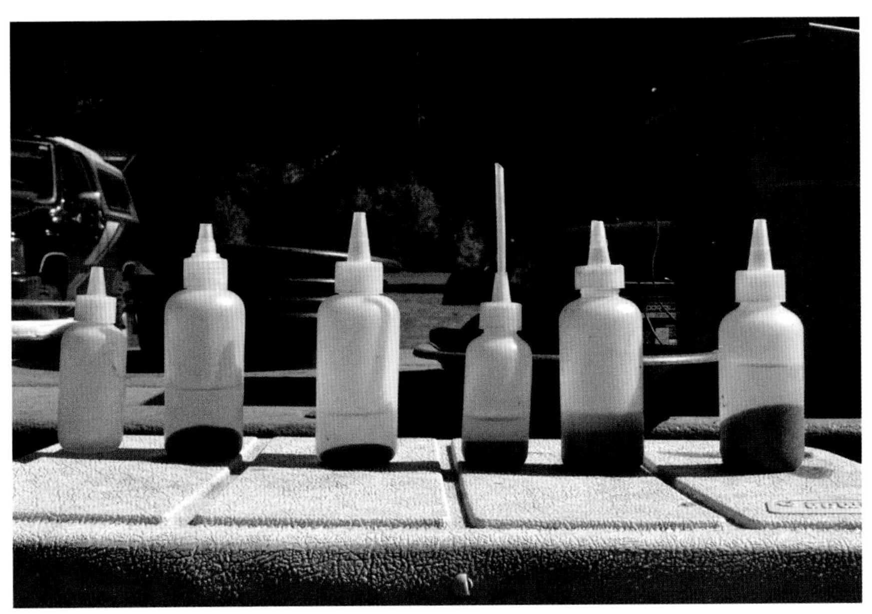

One of the best ways to spend a day prospecting is to hit several different spots and collect GPS coordinates that you can associate with samples. Then clean it all up back in camp and take notes.

of gold. You can easily fill up your small glass specimen jars by removing the straw and emptying the contents of the snuffer bottle; the heavies will concentrate in the tip.

Some folks use an adapted syringe to suck up their concentrates, then dump them into a specimen jar. I've seen someone use a turkey baster, too. When touring lots of places, I'll often just use one snuffer per location, and then figure it all out later. That's where good labeling comes in, and taking notes.

Panning Machines

There are two kinds of panning machines: miniature sluices and spiral pans. Both work with a battery, usually 12-volt, and in a tub, although some versatile systems come with solar panels and work right in the stream. The miniature sluice usually has an upper deck and a lower deck, uses riffles or screens, some carpeting or plastic mini-riffles, and still traps a lot of black sands with the gold. Spiral pans work by washing off lighter material as the spiral brings material to the center, where it can either go through a hole to a little bucket in the back or stay in the middle in a small, removable micro-bucket. The only

Panning machines can take some of the wear and tear off for an "old dog" like my uncle, Doug Romaine.

real trick, usually, is getting the spray right, on a machine such as the Camel, or getting the tilt right, on a machine such as the Gold Magic. I like them both because they have their own tubs, and the only remaining challenge is access to clean water, as you may have to empty and refill your tub every hour. We like to bring multiple buckets and pour the water in to let it settle, or in some places in the Pacific Northwest, we just drain rainwater from our blue tarp into buckets.

The beauty of these machines is that they are easy to set up, somewhat quiet, and if you have a picnic table back at camp, you can gather concentrates all day and pan at night by lantern. Panning machines work with rich concentrates you make from a dredge, sluice, trommel, or highbanker, or they'll also work with the black sands from ten to twenty pans you collected that day. Some machines will even take raw pay dirt, but it's always better to screen the material so you're not dealing with rocks and pebbles.

Sluices

The next step up from panning machines is the sluice, a modern adaptation of an old standby in the goldfields. Sluices are tricky because you have to get both the angle and the water flow correct. If you don't have enough water

This sluice is set up with the right angle and the right water flow to start processing material.

going in, you won't clean your gravels. If you have too much, you can blow even heavies out the end. If you have the water and the angle adjusted correctly, then you're set. You can bring buckets of pay dirt to the sluice and wash a lot of material, and you will have a continual source of clean water, unlike working with panning machines at camp. Sluices usually have three parts—the shell, a set of removable riffles, and a thick artificial turf or carpeting under the riffles. The heavies build up behind the riffles and in the carpeting, so you clean up by emptying material into a bucket and washing out the carpet.

When you have the sluice set up correctly, you don't need to clean up very often. But many crews like to clean up once or twice a day or more, just to make sure they don't lose anything. It's a good idea when you're first starting out to check frequently and compare against what you get with hand-panned samples. That way you can check your recovery. It's also wise to grab a pan of material from right below the end of the sluice on occasion and pan that—just to see if your system is running properly.

Sluicing is a big step up in making concentrates, and you can move a lot of material—5 or 10 times what you can manually pan. At this point you are graduating from prospector to miner, and your goal is to keep feeding your sluice, keep moving material, and keep conducting cleanups. The downside is that many areas are only open to panning, and a sluice is not appropriate. But in many areas where machines with pumps are not allowed, a sluice is still acceptable.

Highbankers are a great way to work gravel bars and islands where a dredge would be impractical.

Once you commit to a dredge, you'll need a team of mining partners, a claim, and enough money to invest in gas, parts, and so on.

Highbanking

The highbanker is a contraption that allows you to work on the banks of a creek or river, rather than in the stream. Where dredges have pontoons and float a sluice that they feed with suction, a highbanker usually works with the pump down at the water, delivering suction to a hole to pull material up to a two-decked sluice. Many times you'll be required to get a permit to run a highbanker, but you can move a lot of material, so it might be worth it.

Dredging

The last small-scale operation that prospectors use to sample ground is the portable dredge. There are backpack dredges to get into remote areas, and there are big 8-inch and 10-inch monsters, or larger, that you'll see on TV on the *Bering Sea Gold* show. There are many more moving parts on a dredge, and a lot more can break down. But suction dredging is perfect for cleaning bedrock and moving lots of material.

Naturally, they're more expensive, but then, you get what you pay for.

Metal Detectors

Using a metal detector takes patience and mechanical aptitude. You need to be willing to invest a few hours into swinging your detector, at minimum, before you can expect to be good at it. You need to dig up all your targets, too, when starting out, so that you can tune your ear to what your detector

If you have the patience and enjoy electronic devices, a metal detector can pay off big in some of the large tailings fields found in the Pacific Northwest. Shown here is a small Falcon gold pinpointer and an older Minelabs device.

is trying to tell you. The best detectors give you a reading so that you can get an idea of what your target might be, but when starting out, you need to dig it up to verify that a nickel reading is actually a nickel. Some hot rocks will confuse your machine, and the basalt found all over Oregon, Washington, and Idaho can be a confounding rock in particular. In many places in the Pacific Northwest, you'll run into massive dredge fields with tailings piles that cover many, many acres. These are great places to hunt for gold nuggets, and you'll probably be removing a lot of nails, wires, screws, and scrap metal too. The payoff can be enormous, and detecting can be a relaxing way to spend a day.

APPENDIX B: BIBLIOGRAPHY

These older books still have good general information and plenty of "how-to" advice on getting started—but they don't have GPS coordinates.

I've been collecting books on Pacific Northwest mining history and gold prospecting for decades. Here are some of my favorite titles, starting with books that cover the region. I've included annotations for that section to help you understand what I am looking for in a good title to add to a personal library.

General

Alt, David, and Donald Hyndman. *Roadside Geology of . . . .* Missoula, MT: Mountain Press Publishing Company.

I'm a big fan of the *Roadside Geology* series. The authors aren't always as sympathetic as I'd like to the hard work put in by the miners of yesteryear, but they do a good job of explaining the general geology of many areas that gold prospectors seek out. They have individual guides for Idaho, Oregon, and Washington.

Bohmker, Tom T. H. *Gold Panner's Guide to the Oregon Cascades.* Independence, OR: Cascade Mountains Gold, 2010.

Tom Bohmker is a legend in Pacific Northwest gold prospecting. He's been a miner, geologist, engineer, mining-supplies shop owner, author, lecturer, and more. He has individual guides for Washington (2009) and Idaho (2012) and detailed guides for the Oregon Cascades (2010), southwest Oregon (2006), eastern Oregon (2011), and the Bohemia District (2005). His books contain lots of stories, anecdotes, and personal adventures, and he has good information about staking claims and the business side of prospecting.

Fisher, Tim. *Ore-Rock-On on DVD, v. xx.* Sandy, OR.

Superb collection of GPS coordinates for rocks, gems, and fossils mostly. Because there is a lot of spillover between minerals and mining, Fisher's DVD comes in very handy.

Gold Prospectors Association of America. *GPAA Claims Club Membership Mining Guide.* Temecula, CA: GPAA.

Best value for prospectors who cover multiple states and want to explore throughout North America. Multiple claims in the Pacific Northwest. Note that just owning the book doesn't make you a club member.

Johnson, Robert Neil. *Gold Diggers Atlas.* Susanville, CA: Cy Johnson & Son, 1971. 64 pp.

An old favorite showing general areas to explore. Very general information, not updated in decades, with questionable accuracy around roads, but includes historic places and won't let you down as far as finding historic districts.

Johnson, Robert Neil. *N.W. Gem Fields and Ghost Town Atlas.* Susanville, CA: Cy Johnson & Son, 1969. 48 pp.

Great guide showing old ghost towns and rockhounding locales in the Pacific Northwest. Most of the ghost towns included in this guide are mining towns.

Koschmann, A. H., and M. H. Bergendahl. Principal Gold-Producing Districts of the United States, US Geological Survey *Professional Paper 610.* Washington, DC: US Government Printing Office, 1968. 283 pp.

An excellent overview of the top gold-producing districts in the United States. Includes production figures up to 1968. Very general as far as placer information and relies on old township/range location information. Excellent research guide for additional USGS reports when looking up specific information about a district. Great to have in PDF format.

Litton, Evie. *Hiking Hot Springs in the Pacific Northwest.* Guilford, CT: FalconGuides, 2001. 339 pp.

After just a few days of prospecting, it's a good idea to get everyone in camp a good soak. This is an excellent guide, specializing in Pacific Northwest locales.

Patera, Alan H. *Pacific Northwest Mining Towns.* Lake Grove, OR: Western Places, 1994. 66 pp.

Good historical information about old mining districts.

Romaine, Garret. *The Modern Rockhounding and Prospecting Handbook*. Guilford, CT: FalconGuides, 2014. 253 pp.

Good overview for prospecting and field geology, with details about reading geology maps, identifying key rocks and minerals, planning successful trips, and more.

US Geological Survey, Mineral Resources of the United States—Calendar Year 1907. Washington, DC: US Government Printing Office, 1908. 743 pp.

Any of these reports are good to have for overview information, even though out-of-date.

Idaho

Adams, Mildrette. *Historic Silver City: The Story of the Owyhees*. Homedale, ID: Owyhee Publishing Co., 1999. 98 pp.

Ballard, Samuel M. *Geology and Gold Resources of Boise Basin, Boise County, Idaho*. Moscow, ID: Idaho Bureau of Mines and Geology, 1924. 103 pp.

Campbell, Arthur. *Thirty-ninth Annual Report of the Mining Industry of Idaho, for the Year 1937*. Boise, ID: State of Idaho, 1938. 309 pp.

Fanselow, Julie. *Idaho Off the Beaten Path*, 6th ed. Guilford, CT: Globe Pequot, 2006. 194 pp.

Hackbarth, Linda. *Bayview and Lakeview, and Other Early Settlements on Southern Lake Pend Oreille before 1940*. Coeur d'Alene, ID: Museum of North Idaho, 2003. 128 pp.

Hendrickson, Borg, and Linwood Laughy. *Clearwater Country! The Traveler's Historical and Recreational Guide*. Kooskia, ID: Mountain Meadow Press 1990. 171 pp.

Raisch, Bruce A. *Ghost Towns of Idaho: The Search for El Dorado*. Virginia Beach, VA: The Donning Co. Publishers, 2008. 160 pp.

Ream, Lanny R. *The Gem & Mineral Collector's Guide to Idaho*. Baldwin Park, CA: Gem Guides Book Co., 2000. 79 pp.

Ream, Lanny R. *Gem Trails of Idaho and Western Montana*. Baldwin Park, CA: Gem Guides Book Co., 2012. 256 pp.

Ream, Lanny R. *Idaho Minerals: The Complete Reference and Guide to the Minerals of Idaho*, 2nd ed. Coeur d'Alene, ID: Museum of Northern Idaho, 2004. 373 pp.

Romaine, Garret. *Rockhounding Idaho*, 2nd ed. Guilford, CT: FalconGuides, 2020. 254 pp.

Sparling, Wayne. *Southern Idaho Ghost Towns*. Caldwell, ID: Caxton Printers, Ltd., 1996. 135 pp.

Welch, Julia Conway. *Gold Town to Ghost Town: The Story of Silver City, Idaho*. Moscow: University of Idaho Press, 1982. 124 pp.

Wells, Merle W. *Gold Camps and Silver Cities: Nineteenth-Century Mining in Central and Southern Idaho*. Moscow: University of Idaho Press, 2002. 233 pp.

Oregon

Baldwin, Ewart M. *Geology of Oregon*. Dubuque, IA: Kendall/Hunt Publishing Co., 1964. 147 pp.

Bishop, Ellen Morris. *In Search of Ancient Oregon*. Portland, OR: Timber Press, 2003. 288 pp.

Bishop, Ellen Morris, and John Eliot Allen. *Hiking Oregon's Geology*. Seattle: The Mountaineers, 1996. 201 pp.

Brooks, Howard. *A Pictorial History of Gold Mining in the Blue Mountains of Eastern Oregon*. Baker, OR: Baker County Historical Society, 2007. 200 pp.

Brooks, Howard, and Len Ramp. *Gold and Silver in Oregon*. Oregon Dept. of Geology and Mineral Industries Bulletin 61, 1968. 336 pp.

Dreisbach, Robert. *Guide to Northeast Oregon*. Seattle: Entropy Conservationists, 1995. 156 pp.

Jackson, Kerby. "Gold at Savage Rapids on the Rogue River." youtube.com/watch?v=u0yCyfPh6Tk. 2010. 8:09.

Johnson, Lars W. *Rockhounding Oregon*. Guilford, CT: FalconGuides, 2014. 280 pp.

May, Keith F. *Ghosts of Times Past: A Road Trip of Eastern Oregon Ghost Towns*. Portland, OR: Drigh Sighed Publications, 1996. 152 pp.

Romaine, Garret. *Gem Trails of Oregon*, 3rd ed. Baldwin Park, CA: Gem Guides, 2008. 272 pp.

Sullivan, William L. *100 Hikes/Travel Guide for Eastern Oregon*. Eugene, OR: Navillus Press, 2001. 240 pp.

Tabor, James Waucop. *Granite and Gold: The Story of Oregon's Smallest City*. Baker, OR: Record-Courier Printers, 1988. 86 pp.

Washington

Amara, Mark S., and George E. Neff. *Geologic Road Trips in Grant County, Washington*. Rochester, WA: Gorham Printing, 2003. 93 pp.

Babcock, Scott, and Bob Carson. *Hiking Washington's Geology*. Seattle: The Mountaineers, 2000. 272 pp.

Bancroft, Howland, and Waldemar Lindgren. *The Ore Deposits of Northeastern Washington*. USGS Bulletin 550, 1914. 215 pp.

Barlee, N. L. *Gold Creeks and Ghost Towns of Northeastern Washington*. Surrey, BC: Hancock House Publishers, 1999. 223 pp.

Cannon, Bart. *Minerals of Washington*. Mercer Island, WA: Cordilleran Press, 1975. 187 pp.

Jackson, Bob. *The Panner's Guide to Northwest Gold*. Renton, WA: Bob Jackson Books, 1980. 46 pp.

Mayo, Roy F. *Washington State Gold Mines*. Wenatchee, WA: Nugget Enterprises, 1983. 80 pp.

Moen, Wayne S. *Myers Creek and Wauconda Mining Districts of Northeastern Okanogan County, Washington*. Washington Dept. of Natural Resources Bulletin 73, 1980. 96 pp.

Moen, Wayne S. *Silver Occurrences of Washington*. Washington Dept. of Natural Resources Bulletin 69, 1976. 188 pp.

Northwest Underground Explorations. *Discovering Washington's Historic Mines, Volume 1: The West Central Cascade Mountains*. Arlington, WA: Oso Publishing Co., 1997. 230 pp.

Northwest Underground Explorations. *Discovering Washington's Historic Mines, Volume 2: The East Central Cascade Mountains and the Wenatchee Mountains*. Arlington, WA: Oso Publishing Co., 2002. 336 pp.

Northwest Underground Explorations. *Discovering Washington's Historic Mines, Volume 3: The Northern Cascade Mountains*. Arlington, WA: Oso Publishing Co., 2006. 315 pp.

Northwest Underground Explorations. *Discovering Washington's Historic Mines, Volume 4: The Western Okanogan*. Arlington, WA: Oso Publishing Co., 2011. 399 pp.

Ream, Lanny. *Gems and Minerals of Washington*, 3rd ed. Renton, WA: Jackson Mountain Press, 1994. 216 pp.

Romaine, Garret. *Gem Trails of Washington*, 2nd ed. Baldwin Park, CA: Gem Guides, 2014. 200 pp.

Smith, Jerry. *Boom Towns and Relic Hunters of Northeastern Washington*. Bellevue, WA: Elfin Cove Press, 2002. 124 pp.

Washington Geology, vol. 24, No. 2, "Republic Centennial Issue." Washington Dept. of Natural Resources, June 1996. 44 pp.

APPENDIX C: WEBSITES

I've been tracking gold-prospecting links on the internet since 1996 and wrote a regular column for the GPAA magazine titled "Mining the Internet." If there's one thing you can count on, it's the fact that things change quickly in cyberspace. These are great places to begin your online research, but your web sleuthing could turn up more good information.

General

Bureau of Land Management (BLM)
blm.gov/wo/st/en.html
Link to local offices for each state.

US Forest Service
fs.usda.gov
Good starting point to find the local districts and forest managers for areas you plan to visit.

US Geological Survey
usgs.gov
Learn the science behind gold prospecting.

Gold Prospecting Online
goldprospectingonline.com
Good site for general gold-mining information; good info for beginners, and sells equipment.

Google Earth
One of my favorite programs for tracking where I've been. I have a GPS device that I can synch with Google Earth and import my waypoints and tracks, which sometimes shows how close I got to a place I had my heart set on visiting.

Mine Cache

This overlay for Google Earth places icons for gold mines and major prospects into Google Earth. It's not that spendy at $29.95 per year, and although it's not verified in the field, it gives you a good idea of what major mine is nearby.

Mindat

mindat.org

Primarily a site for mineral collectors, this extensive database has good information for gold prospectors too. Here's a search for gold in Oregon, which brings up nineteen pages of information down to the GPS coordinates. You could substitute Washington or Idaho: www.mindat.org/minlocsearch.php?frm_id=mls&cform_is_valid=1&cf_mls_page=1&minname=gold®ion=Oregon

Mineral Resources Data System (MRDS)

mrdata.usgs.gov/mrds

An excellent source of information packed by the US Geological Survey, with detailed information about the geology of significant mines in an area. You can export information to build your own spreadsheet as well. One downside is that the general map includes sand and gravel operations, so it helps to sort by commodity before starting. The report also often describes which particular references include more information, and with more and more of these coming online, your online research promises to yield better and better results.

Prospectors Paradise

prospectorsparadise.com

General info about gold mining and an active forum for asking questions.

Water levels (USGS)

waterdata.usgs.gov/id/nwis/rt

Good for determining how high the water is at a stream or river you plan to visit.

Western Mining History

westernmininghistory.com/mine_detail/10047836

Good background information about the major gold-mining districts in the West.

Washington

Department of Fish and Wildlife

wdfw.wa.gov/licensing/mining

Permits and links to the "Gold and Fish" pamphlet, which you should have on you at all times when walking Washington streams.

Department of Natural Resources

dnr.wa.gov/RecreationEducation/Topics/HarvestingCollecting/Pages/mineral_collecting.aspx

Good springboard for more DNR information.

Olympics Gold

olygold.com/findgold.html

All about gold prospecting in the Olympics.

Water Access

wa.gov/lands/water_access

Fast access to information about which creeks, streams, rivers, and lakes have public access.

Oregon

Department of Oregon Geology and Mineral Industries

oregon.gov/DOGAMI/Pages/index.aspx

Good starting point for tracking down Oregon's historic mines and districts.

Gold in Oregon

goldrushnuggets.com/goldinoregon.html

Overview, how-to, and more.

Kerby Jackson's blog

kerbyjackson.com

Great source for history, background, and access to this prolific writer's newest work.

Nature of the Northwest (DOGAMI store)

naturenw.org/recreation.htm

Good source for permits, access to old geology reports.

New 49ers
goldgold.com/tag/gold-mining-oregon
Awesome club run by Dave McCracken, a legend in gold prospecting and mining circles.

Oregon Gold
oregongold.net
Excellent site with plenty of information about known gold-bearing districts. Good how-to articles as well.

Photograph Oregon
photographoregon.com/Oregon-Mines.html
Good shots of what's still standing out in the hills.

Rogue River Information
blm.gov/or/resources/recreation/rogue/gold-panning.php

Idaho

Bedrock or Bust
bedrockorbust.blogspot.com
Good blog from Bedrock Bob.

Crystal Gold Mine Tour
goldmine-idaho.com

Idaho Department of Lands
idl.idaho.gov/bureau/Minerals/gem_guide/gg_index.htm

Gold Mining Stories from Murray
murray-idaho.com

Idaho Geological Survey Geology Maps
idahogeology.org/Products/reverselook.asp?switch=pubs&value=Digital_Web_Maps_(DWM)

Idaho Geological Survey Mines and Minerals Database
idahogeology.org/Services/MinesAndMinerals/default.htm

Idaho Gold and Gem Outfitters
idahogoldandgemoutfitters.com
Excellent site with lots of information about Idaho.

Idaho Museum of Mining and Geology
idahomuseum.org

Kellogg, Idaho, Mining District
murray-idaho.com

Minerals of Latah County
webpages.uidaho.edu/~mgunter/geol249/latah/location_index.html

Mining in Idaho
imnh.isu.edu/digitalatlas/geog/mining/minemain.htm

Pierce Chamber of Commerce
pierce-weippechamber.com

State Land
parksandrecreation.idaho.gov

APPENDIX D: CLUBS AND ORGANIZATIONS

I cannot stress enough the importance of joining a gold-prospecting club. Not only do you get access to valid claims, you also begin building a network of contacts with knowledge about prospecting, equipment, and regional hot spots by attending regular club meetings and annual shows. Some clubs even have equipment you can borrow or rent, and some bring equipment to meetings so you can practice, clean up concentrates, and so forth.

The downside is that it would be nice to get more reciprocal benefits, especially for those of us who like to roam from state to state, or even from district to district. It is annoying to hear, "You're in the wrong club!"

Below are some links to get you started. I did my best to update everything and scoured the web for clubs I didn't know about, but it's possible I missed some. More groups are relying on Facebook to host their web presence, so be sure to try some searches there as well.

National

Gold Miners Headquarters
goldminershq.com/clubs/gold2.htm
Contains a list of clubs by state.

Gold Prospectors Association of America (GPAA)
goldprospectors.org
My personal favorite, with claims in all western states and a large, active reach. They have annual shows in several US cities, and an excellent magazine that I've written for since 1996.

Prospecting Channel
prospectingchannel.com/indexClub.html
Good updated list for clubs both national and worldwide.

Regional

Western Mining Alliance
westernminingalliance.org
Large organization of independent miners.

Washington

Bedrock Prospectors Club
bedrockprospectors.org
Active club in Puyallup with claims and outings. Good links to mining reports.

North American Miners Association
blackjacksmetaldetectors.com/#!north-american-miners-association/c1wnq
Based in Renton, Washington. Links to other metal-detecting clubs.

North American Prospectors
auor.tripod.com
Based in Chehalis.

North Central Washington Prospectors
facebook.com/ncwp.goldclub
Located in Wenatchee.

Northwest Mineral Prospector Club
(See Oregon listing.)

Prospectors Plus
prospectorsplus.com
Active club headquartered at store in Gold Bar with claims near Index, Galena, Liberty, and elsewhere.

Washington Prospectors Mining Association
washingtonprospectors.org
Very active club with claims especially on Sultan River, but also near Index and elsewhere, with outings and annual shows.

Oregon

Bohemia Mine Owners Association
bohemiamineownersassociation.webs.com
Concentrated near Cottage Grove, Oregon, and specializing in the Bohemia area.

Douglas County Prospectors
dcpagold.com
Small club in southwest Oregon gold country.

Eastern Oregon Miners and Prospectors
eomp.org
Located in Baker City, with good claims near John Day, Granite, and upper Burnt River, among others.

Eastern Oregon Mining Association
h2oaccess.com
Headquartered in Baker City, Oregon. Does not own any claims, but dedicated to gold mining.

Gold Coast Mining Association
No website. Try oregongoldhunters.com/index.php.

Jefferson Mining District
jeffersonminingdistrict.com
Dedicated to preserving land rights for prospectors, especially in southern Oregon and northern California.

Josephine County Sourdoughs
No website. Try oregongoldhunters.com/index.php.

Millennium Diggers Club
millenniumdiggers.com
Based in Keizer, Oregon. All buried treasure, be it rocks, gems, artifacts, and so forth.

Northwest Mineral Prospector Club (NWMPC)
nwmpc.com/joomla
Active club with claims in Oregon at Molalla, Quartzville Creek, Brice Creek, Chetco River, and Coquille River, and Copper Creek in Washington.

Oregon Independent Miners
No website.

Willamette Valley Miners
wvminers.com
Active club with claims in Oregon.

Idaho

Idaho GPAA
idahogpaa.com
Nampa club that's part of national organization.

Northwest Gold Prospectors Association
icehouse.net/blowe/nwgold.html or facebook.com/nwgpaclearwaterchapter
Headquartered in northern Idaho.

Northwest Gold Prospectors
icehouse.net/blowe/nwgold.html
Active club with summer outings.

INDEX

ABOUT THE AUTHOR

Garret Romaine is an avid gold prospector, rockhound, and fossil collector with years of experience in the field. He is a longtime writer for *Gold Prospectors* magazine and is the author of *The Modern Rockhounding and Prospecting Handbook*; *Rocks, Gems, and Minerals of the Southwest*; *Rocks, Gems, and Minerals of the Rocky Mountains*; and *Rockhounding Idaho*, all from FalconGuides, as well as *Gem Trails of Washington* and *Gem Trails of Oregon*. Garret is a former executive director of the Rice Northwest Museum of Rocks and Minerals in Hillsboro, Oregon.